3-D Shape Estimation and Image Restoration

Paolo Favaro and Stefano Soatto

3-D Shape Estimation and Image Restoration

Exploiting Defocus and Motion Blur

 Springer

Paolo Favaro
Heriot-Watt University, Edinburgh, UK
http://www.eps.hw.ac.uk/~pf21

Stefano Soatto
University of California, Los Angeles, USA
http://www.cs.ucla.edu/~soatto

British Library Cataloguing in Publication Data
A catalogue record for this book is available from the British Library

Library of Congress Control Number: 2006931781

Printed on acid-free paper
ISBN 978-1-84996-559-0

9 8 7 6 5 4 3 2 1

Springer Science+Business Media
springer.com

To Maria and Giorgio

Paolo Favaro

To Anna and Arturo

Stefano Soatto

Preface

Images contain information about the spatial properties of the scene they depict. When coupled with suitable assumptions, images can be used to infer three-dimensional information. For instance, if the scene contains objects made with homogeneous material, such as marble, variations in image intensity can be associated with variations in shape, and hence the "shading" in the image can be exploited to infer the "shape" of the scene (shape from shading). Similarly, if the scene contains (statistically) regular structures, variations in image intensity can be used to infer shape (shape from textures). Shading, texture, cast shadows, occluding boundaries are all "cues" that can be exploited to infer spatial properties of the scene from a single image, when the underlying assumptions are satisfied. In addition, one can obtain spatial cues from multiple images of the same scene taken with changing conditions. For instance, changes in the image due to a moving light source are used in "photometric stereo," changes in the image due to changes in the position of the cameras are used in "stereo," "structure from motion," and "motion blur." Finally, changes in the image due to changes in the geometry of the camera are used in "shape from defocus." In this book, we will concentrate on the latter two approaches, motion blur and defocus, which are referred to collectively as "accommodation cues." Accommodation cues can be exploited to infer the 3-D structure of the scene as well as its radiance properties, which in turn can be used to generate better quality novel images than the originals.

Among visual cues, defocus has received relatively little attention in the literature. This is due in part to the difficulty in exploiting accommodation cues: the mathematical tools necessary to analyze accommodation cues involve continuous analysis; unlike stereo and motion which can be attacked with simple linear algebra. Similarly, the design of algorithms to estimate 3-D geometry from accommodation cues is more difficult because one has to solve optimization problems in infinite-dimensional spaces. Most of the resulting algorithms are known to be slow and lack robustness in respect to noise.

Recently, however, it has been shown that by exploiting the mathematical structure of the problem one can reduce it to linear algebra, (as we show in Chapter 4,) yielding very simple algorithms that can be implemented in a few lines of code. Furthermore, links established with recent developments in variational methods allow the design of computationally efficient algorithms. Robustness to noise has significantly improved as a result of designing optimal algorithms.

This book presents a coherent analytical framework for the analysis and design of algorithms to estimate 3-D shape from defocused and blurred images, and to eliminate defocus and blur and thus yield "restored" images. It presents a collection of algorithms that are shown to be optimal with respect to the chosen model and estimation criterion. Such algorithms are reported in MATLAB® notation in the appendix, and their performance is tested experimentally.

The style of the book is tailored to individuals with a background in engineering, science, or mathematics, and is meant to be accessible to first-year graduate students or anyone with a degree that included basic linear algebra and calculus courses. We provide the necessary background in optimization and partial differential equations in a series of appendices.

The research leading to this book was made possible by the generous support of our funding agencies and their program managers. We owe our gratitude in particular to Belinda King, Sharon Heise, and Fariba Fahroo of AFOSR, and Behzad Kamgar-Parsi of ONR. We also wish to thank Jean-Yves Bouguet of Intel, Shree K. Nayar of Columbia University, New York, and also the National Science Foundation.

September 2006   PAOLO FAVARO
                                                       STEFANO SOATTO

Organizational chart

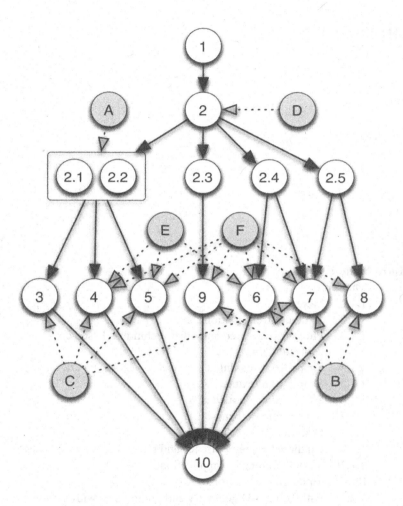

Figure 1. Dependencies among chapters.

Contents

1
Introduction

1.1 The sense of vision

The sense of vision plays an important role in the life of primates, by facilitating interactions with the environment that are crucial for survival tasks. Even relatively "unintelligent" animals can easily navigate through unknown, complex, dynamic environments, avoid obstacles, and recognize prey or predators at a distance. Skilled humans can view a scene and reproduce a model of it that captures its shape (sculpture) and appearance (painting) rather accurately.

The goal of any visual system, whether natural or artificial, is to infer properties of the environment (the "scene") from images of it. Despite the apparent ease with which we interact with the environment, the task is phenomenally difficult, and indeed a significant portion of the cerebral cortex is devoted to it: [Felleman and van Essen, 1991] estimate that nearly half of the cortex of macaque monkeys is engaged in processing visual information. In fact, visual inference is strictly speaking impossible, because the complexity of the scene is infinitely greater than the complexity of the images, and therefore one can never hope to recover "the" correct model of the scene, but only a representation of it. In other words, visual inference is an ill-posed inverse problem, and therefore one needs to impose additional structure, or make additional assumptions on the unknowns.

For the sake of example, consider an image of an object, even an unfamiliar one such as that depicted in Figure 1.1. As we show in Chapter 2, an image is generated by light reflected by the surface of objects in ways that depend upon their material properties, their shape, and the light source distribution. Given an image, one can easily show that there are infinitely many objects which have different

Figure 1.1. An image of an unfamiliar object (image kindly provided by Silvio Savarese). Despite the unusual nature of the scene, interpretation by humans is quite consistent, which indicates that additional assumptions or prior models are used in visual inference.

shape, different material properties, and infinitely many light source distributions that generate that particular image. Therefore, in the absence of additional information, one can never recover the shape and material properties of the scene from the image. For instance, the image in Figure 1.1 could be generated by a convex jar like 3-D object with legs, illuminated from the top, and viewed from the side, or by a flat object (the surface of the page of this book) illuminated by ambient light, and viewed head-on. Of course, any combination of these two interpretations is possible, as well as many more. But, somehow, interpretation of images by humans, despite the intrinsic ambiguity, is remarkably consistent, which indicates that strong assumptions or prior knowledge are used. In addition, if more images of the object are given, for instance, by changing the vantage point, one could rule out ambiguous solutions; for instance, one could clearly distinguish the jar from a picture of the jar.

Which assumptions to use, what measurements to take, and how to acquire prior knowledge, is beyond the realm of mathematical analysis. Rather, it is a *modeling task*, which is a form of engineering art that draws inspiration from studying the mathematical structure of the problem, as well as from observing the behavior of existing visual systems in biology. Indeed, the apparent paradox of the prowess of the human visual system in face of an impossible task is one of the great scientific challenges of the century.

Although "generic" visual inference is ill-posed, the task of inferring properties of the environment may become well-posed within the context of a specific task. For instance, if I am standing inside a room, by vision alone I cannot recover the correct model of the room, including the right distances and angles. However, I can recover a model that is good enough for me to move about the room without hitting objects within. Or, similarly, I can recover a model that is good enough for me to depict the room on a canvas, or to reproduce a scaled model of it. Also, just a cursory look is sufficient for me to develop a model that is sufficient for me to recognize this particular room if I am later shown a picture of it. In fact, because visual inference is ill-posed, the choice of representation becomes critical, and the task may dictate the representation to use. It has to be rich enough to allow accomplishing the task, and yet exclude all "nuisance factors" that affect the mea-surements (the image) but that are irrelevant to the task. For instance, when I am driving down the road, I do not need to accurately infer the material properties of buildings surrounding it. I just need to be able to estimate their position and a rough outline of their shape.

Tasks that are enabled by visual inference can be lumped into four classes that we like to call reconstruction, reprojection, recognition, and regulation. In re-construction we are interested in using images to infer a spatial model of the environment. This could be done for its own sake (for instance, in sculpture), or in partial support of other tasks (for instance, recognition and regulation). In re-projection, or rendering, we are interested in using images to infer a model that allows one to render the scene from a different viewpoint, or under different illu-mination. In recognition, or more in general in classification and categorization, we are interested in using images to infer a model of objects or scenes so that we can recognize them or cluster them into groups, or more in general make decisions about their identity. For instance, after seeing a mug, we can easily rec-ognize that particular mug, or we can simply recognize the fact that it is a mug. In regulation we are interested in using vision as a sensor for real-time control and interaction, for instance, in navigating within the environment, tracking objects, grasping them, and so on.

In this book we concentrate on reconstruction tasks, specifically on the esti-mation of 3-D shape, and on rendering tasks, in particular image restoration.[1] At

---

[1]The term "image restoration" is common, but misleading. In fact, our goal is, from a collection of images taken by a camera with a finite aperture, to recover the 3-D shape of the scene and its "radiance." We will define radiance properly in the next chapter, but for now it suffices to say that the radiance is a property of the scene, not the image. The radiance can be thought of as the "ideal"

this level of generality these tasks are not sufficient to force a unique representation and yield a well-posed inference problem, so we need to make additional assumptions on the scene and/or on the imaging process. Such assumptions result in different "cues" that can be studied separately in order to unravel the mathematical properties of the problem of visual reconstruction and reprojection. Familiar cues that the human visual system is known to exploit, and that have been studied in the literature, are stereo, motion, texture, shading, and the like, which we briefly discuss below. In order to make the discussion more specific, without needing the technical notation that we have yet to introduce, we define the notion of "reflectance" informally as material properties of objects that affect their interaction with light. Matte objects such as unpolished stone and chalk exhibit "diffuse reflectance" in the sense that they scatter light in equal amounts in all directions, so that their appearance does not change depending on the vantage point. Shiny objects such as plastic, metal, and glass, exhibit "specular reflectance," and their appearance can change drastically with the viewpoint and with changes in illumination.

The next few paragraphs illustrate various visual cues, and the associated assumptions in reflectance, illumination, and imaging conditions.

1.1.1 Stereo

In stereo one is given two or more images of a scene taken from different vantage points. Although the relative position and orientation of the cameras are usually known through a "calibration" procedure, this assumption can be relaxed, as we discuss in the next paragraph on structure from motion. In order to establish "correspondence" among images taken from different viewpoints, it is necessary to assume that the illumination is constant, and that the scene exhibits diffuse reflection. Barring these assumptions, one can make images taken from different vantage points arbitrarily different by changing the illumination. Consider, for instance, two arbitrarily different images. One can build a scene made from a mirror sphere, and two illumination conditions obtained by back-projecting the images through the mirror sphere onto a larger sphere. Naturally, it would be impossible to establish correspondence between these two images, although they are technically portraying the same scene (the mirror sphere). In addition, even if the scene is matte, in order to establish correspondence we must require that its reflectance (albedo) be nowhere constant. If we look at a white marble sphere on a white background with uniform diffuse illumination, we will see white no matter what the viewpoint is, and we will be able to say nothing about the scene! Under these assumptions, the problem of reconstructing 3-D shape is well understood because it reduces to a purely geometric construction, and several textbooks have addressed

---

or "restored" or "deblurred" image, but in fact it is much more, because it also allows us to generate novel images from different vantage points and different imaging settings.

this task; see for instance, [Ma et al., 2003] and references therein. Once 3-D shape has been reconstructed, reprojection, or reconstruction of the reflectance of the scene, is trivial. In fact, it can be shown that the diffuse reflectance assumption is precisely what allows one to separate the estimate of shape (reconstruction) from the estimate of albedo (reprojection) [Soatto et al., 2003].

More recent work in the literature has been directed at relaxing some of these assumptions: it is possible to consider reconstruction for scenes that exhibit diffuse + specular reflection [Jin et al., 2003a] using an explicit model of the shape and reflectance of the scene. Additional reconstruction schemes either model reflectance explicitly or exhibit robustness to deviations of the diffuse reflectance assumption [Yu et al., 2004], [Bhat and Nayar, 1995], [Blake, 1985], [Brelstaff and Blake, 1988], [Nayar et al., 1993], [Okutomi and Kanade, 1993].

1.1.2   Structure from motion

Structure from motion refers to the problem of recovering the 3-D geometry of the scene as well as its motion relative to the camera, from a sequence of images. This is very similar to the multiview stereo problem, except that the mutual position and orientation of the cameras are not known. Unlike stereo, in structure from motion one can often assume the fact that images are taken from a continuously moving camera (or a moving scene), and such temporal coherence has to be taken into account in the inference process. This, together with techniques for recovering the internal geometry of the camera, is well understood and has become commonplace in computer vision (see [Ma et al., 2003] and references therein).

1.1.3   Photometric stereo and other techniques based on controlled light

Unlike stereo, where the light is constant and the viewpoint changes, photometric stereo works under the assumption that the viewpoint is constant, and the light changes. One obtains a number of images of the scene from a static camera after changing the illumination conditions. Given enough images with enough different light configurations, one can recover the shape and also the reflectance of the scene [Ikeuchi, 1981].

It has been shown that if the reflectance is diffuse, one can capture most of the variability of the images using low-dimensional linear subspaces of the space of all possible images [Belhumeur and Kriegman, 1998]. Under these conditions, one can also allow changes in viewpoint, and show that the scene can be recovered up to an ambiguity that affects the shape and the position of the light source [Yuille et al., 2003].

1.1.4   Shape from shading

Although the cues described so far require two or more images with changing conditions being available, in shape from shading one only requires one image of a scene. Naturally, the assumptions have to be more stringent, and in particular one typically requires that the reflectance be diffuse and constant, and that the position of the light source be known (see [Horn and Brooks, 1989; Prados and Faugeras, 2003] and references therein). It is also possible to relax the assumption of diffuse and constant reflectance as done in [Ahmed and Farag, 2006], or to relax the assumption of known illumination by considering multiple images taken from a changing viewpoint [Jin et al., 2003b].

Because reflectance is constant, it is characterized by only one number, so the scene is completely described by the reconstruction process, and reprojection is straightforward. Indeed, shading is one of the simplest and most common techniques to visualize 3-D surfaces in single 2-D images.

1.1.5   Shape from texture

Like shading, texture is a cue that allows inference from a single image. Rather than assuming that the reflectance of the scene is constant, one assumes that certain statistics of the reflectance are constant, which is commonly referred to as *texture-stationarity* [Forsyth, 2002]. For instance, one can assume that the response of a certain bank of filters, which indicate fine structure in the scene, is constant. If the structure of the appearance of the scene is constant, its variations on the image can be attributed to the shape of the scene, and therefore be exploited for reconstruction. Naturally, if the underlying assumptions are not satisfied, and the structure of the scene is not symmetric, or repeated regularly, the resulting inference will be incorrect. This is true of all visual cues when the underlying assumptions are violated.

1.1.6   Shape from silhouettes

In shape from silhouettes one exploits the change of the image of the occluding contours of object. In this case, one must have multiple images obtained from different vantage points, and the reflectance must be such that it is possible, or easy, to identify the occluding boundaries of the scene from the image. One such case is when the reflectance is constant, or smooth, which yields images that are piecewise constant or smooth, where the discontinuities are the occluding boundaries [Cipolla and Blake, 1992; Yezzi and Soatto, 2003]. In this case, shape and reflectance can be reconstructed simultaneously, as shown in [Jin et al., 2000].

1.1.7   Shape from defocus

In the cues described above, multiple images were obtained by changing the position and orientation of the imaging device (multiple vantage points). Alternatively, one could consider changing the geometr, rather than the location, of the imaging device. This yields so-called *accommodation cues.*

When we consider a constant viewpoint and illumination, and collect multiple images where we change, for instance, the position of the imaging sensor within the camera, or the aperture or focus of the lens, we obtain different images of the same scene that contain different amounts of "blur." Because there is no change in viewpoint, we are not restricted to diffuse reflectance, although one could consider slight variations in appearance from different vantage points on the spatial extent of the lens.

In this case, as we show, one can estimate both the shape and the reflectance of the scene [Pentland, 1987], [Subbarao and Gurumoorthy, 1988], [Pentland et al., 1989], [Nayar and Nakagawa, 1990], [Ens and Lawrence, 1993], [Schechner and Kiryati, 1993], [Xiong and Shafer, 1993], [Noguchi and Nayar, 1994], [Pentland et al., 1994], [Gokstorp, 1994], [Schneider et al., 1994], [Xiong and Shafer, 1995], [Marshall et al., 1996], [Watanabe and Nayar, 1996a], [Rajagopalan and Chaudhuri, 1997], [Rajagopalan and Chaudhuri, 1998], [Asada et al., 1998a], [Watanabe and Nayar, 1998], [Chaudhuri and Rajagopalan, 1999], [Favaro and Soatto, 2000], [Soatto and Favaro, 2000], [Ziou and Deschenes, 2001], [Favaro and Soatto, 2002], [Jin and Favaro, 2002], [Favaro et al., 2003], [Favaro and Soatto, 2003], [Rajagopalan et al., 2004], [Favaro and Soatto, 2005]. The latter can be used to generate novel images, and in particular "deblurred" versions of the original ones.

Additional applications of the ideas used in shape from defocus include confocal microscopy [Ancin et al., 1996], [Levoy et al., 2006] as well as recent efforts to build multicamera arrays [Levoy et al., 2004].

1.1.8 Motion blur

All the cues above assume that each image is obtained with an infinitesimally small exposure time. However, in practice images are obtained by integrating energy over a finite spatial (pixel area) and temporal (exposure time) window [Brostow and Essa, 2001], [Kubota and Aizawa, 2002]. When the aperture is open for a finite amount of time, the energy is averaged, and therefore objects moving at different speeds result in different amounts of "blur." The analysis of blurred images allows us to recover spatial properties of the scene, such as shape and motion, under the assumption of diffuse reflection [Ma and Olsen, 1990], [Chen et al., 1996], [Tull and Katsaggelos, 1996], [Hammett et al., 1998], [Yitzhaky et al., 1998], [Borman and Stevenson, 1998], [You and Kaveh, 1999] [Kang et al., 1999], [Rav-Acha and Peleg, 2000], [Kang et al., 2001], [Zomet et al., 2001], [Kim et al., 2002], [Ben-Ezra and Nayar, 2003], [Favaro et al., 2004], [Favaro and Soatto, 2004], [Jin et al., 2005].

1.1.9 On the relative importance and integration of visual cues

The issue of how the human visual system exploits different cues has received a considerable amount of attention in the literature of psychophysics and physiology [Marshall et al., 1996], [Kotulak and Morse, 1994], [Flitcroft et al., 1992], [Flitcroft and Morley, 1997], [Walsh and Charman, 1988]. The gist of this literature is that motion is the "strongest" cue, whereas stereo is exploited far less than commonly believed [Marshall et al., 1996], and accommodation as a cue decreases with age as the muscles that control the shape of the lens stiffen. There are also interesting studies on how various cues are weighted when they are conflicting, for instance, vergence (stereo) and accommodation [Howard and Rogers, 1995]. However, all these studies indicate the relative importance and integration of visual cues for the very special case of the human visual system, with all its constraints on how visual data are captured (anatomy and physiology of the eye) and processed (structure and processing architecture of the brain).

Obviously, any engineering system aiming at performing reconstruction and reprojection tasks will eventually have to negotiate and integrate all different cues. Indeed, depending on the constraints on the imaging apparatus and the application target, various cues will play different roles, and some of them may be more important than others in different scenarios. For instance, in the reconstruction of common objects such as mugs or faces, multiview stereo can play a relatively important role, because one can conceive of a carefully calibrated system yielding high accuracy in the reconstruction. In fact, similar systems are currently employed in high-accuracy quality control of part shape in automotive manufacturing. However, in the reconstruction of spatial structures in small spaces, such as cavities or in endoscopic procedures, one cannot deploy a fully calibrated multiview stereo system, but one can employ a finite-aperture endoscope and therefore exploit accommodation cues. In reconstruction of large-scale structures, such as architecture, neither accommodation nor stereo provides sufficient baseline (vantage point) variation to yield accurate reconstruction, and therefore motion will be the primary cue.

1.1.10   Visual inference in applications

Eventually, we envision engineering systems employing all cues to interact intelligently with complex, uncertain, and dynamic environments, including humans and other engineering systems. To gauge the potential of vision as a sensor, think of spending a day with your eyes closed, and all the things you would not be able to do: drive your car, recognize familiar objects at a distance, grasp objects in one shot, or locate a place or an object in a cluttered scene. Now, think of how engineering systems with visual capabilities can enable a sense of vision for those who have lost it[2] as well as provide support and aid to the visually impaired, or simply relieve us from performing tedious or dangerous tasks, such as driving our cars through stop-and-go traffic, inspecting an underwater platform, or exploring planets and asteroids.

---

[2]Recent statistics indicate that blindness is increasing at epidemic rates, mostly due to the increased longevity in the industrialized world.

The potential of vision as a sensor for engineering systems is staggering, however, most of its potential has been untapped so far. In fact, one could argue that some of the goals set forth above were already enunciated by Norbert Wiener over half a century ago, and yet we do not see robotic partners helping with domestic chores, or even driving us around. This, however, may soon be changing. Part of the reason for the slow progress is due to the fact that, until just over a decade ago, it was simply not possible to buy hardware that could bring a full-resolution image into the memory of a commercial personal computer in real-time (at 30 frames per second), let alone process it and do anything useful with the results. This is no longer the case now, however, and several vision-based systems have already been deployed for autonomous driving on the Autobahn [Dickmanns and Graefe, 1988], autonomous landing on Mars [Cheng et al., 2005], and for automated analysis of movies [web link, 2006a] and architectural scenes [web link, 2006b].

What has changed in the past decade, however, is not just the speed of hardware in personal computers. Early efforts in the study of vision had greatly underestimated the difficulty of the problem from a purely analytical standpoint. It is only in the last few years that sophisticated mathematical tools have been brought to bear to address some of these difficulties, ranging from differential and algebraic geometry to functional analysis, stochastic process, and statistical inference. At this stage, therefore, it is crucial to understand the problem from a mathematical and engineering standpoint, and identify the role of various assumptions, their necessity, and their impact in the well-posedness of the problem. Therefore, a mathematical and engineering analysis of each visual cue in isolation is worthwhile before we venture into meaningfully combining them towards building a comprehensive vision system.

1.2   Preview of coming attractions

In this section we summarize the content of the book in one chapter. The purpose is mostly tutorial, as we wish to give a bird's-eye view of the topic and give some context that will help the reader work through the chapters.

1.2.1   Estimating 3-D geometry and photometry with a finite aperture

The general goal of visual inference is to provide estimates of properties of the scene from images. In particular, we are interested in inferring geometric (shape) and photometric (reflectance) properties of the scene. This is one of the primary goals of the field of computer vision. Most of the literature, however, assumes a simplified model of image formation where the cameras have an infinitesimal aperture (the "pinhole" camera) and an infinitesimal exposure time. The first goal in this book is to establish that having an explicit model of a finite aperture and a finite exposure time not only makes the resulting models more realistic, but also provides direct information on the geometry and photometry of the environment.

In plain terms, the goal of this book can be stated as follows.

> Given a number of images obtained by a camera with a finite aperture and a finite exposure time, where each image is obtained with a different imaging setting (e.g., different aperture, different exposure time, or different focal length, position of the imaging sensor, etc.), infer an estimate of the 3-D structure of the scene and the reflectance of the scene.

Because there is no change in viewpoint, the 3-D structure of the scene can be represented as a depth map, and the reflectance can be represented by the radiance of the scene.

In Chapter 2 we establish the notation and terminology that is used throughout the book. We define precisely what it means to "infer" shape and reflectance, and how shape and reflectance are represented, and how reflectance, radiance, and illumination are related to one another. As we show for the most part these definitions entail modeling choices: we seek a model that is rich enough to capture the phenomenology of the image formation process, and simple enough to allow analysis and design of inference algorithms.

Given a model, it is important to analyze its mathematical properties and the extent to which it allows inference of the desired unknowns, that is, shape and reflectance. However, because the choice of model is part of the design process, its power will eventually have to be validated experimentally. In Chapter 3, which we summarize next, we address the analysis of the mathematical properties of various models, to gauge to what extent this model is rich enough to allow "observation" of the hidden causes of the image formation process. The following chapters address the design of inference algorithms for various choices of models and inference criteria. Each chapter includes experimental evaluations to assess the power of each model and the resulting inference algorithm.

1.2.2   Testing the power and limits of models for accommodation cues

Because visual inference is the "inverse problem" of image formation, after the modeling exercise of Chapter 2, where we derive various models based on common assumptions and simplifications, we must analyze these models to determine under what condition they can be "inverted." In other words, for a given mathematical model, we must study the conditions that allow us to infer 3-D shape and radiance. Chapter 3 addresses this issue, by studying the observability of shape and reflectance from blurred images. There, we conclude that if one is allowed to control the radiance of the scene, for instance, by using structured light, then the shape of any reasonable scene can be recovered. If one is not allowed to control the scene, which is the case of interest in this book, then the extent to which one can recover the 3-D structure of the scene depends on its reflectance. More specifically, the more "complex" the radiance, the more "precise" in the reconstruction. In the limit where the radiance is constant, nothing can be said about 3-D structure (images of a white scene are white no matter what the shape of the scene is).

This analysis serves as a foundation to understand the performance of the algorithms developed later. One can only expect such algorithms to perform under the conditions for which they are designed. Testing a shape from defocus algorithm on a scene with smooth or constant radiance would yield to large error through no fault of the algorithm, because the mathematics reveals that no accurate reconstruction can be achieved, no matter what the criterion and what the implementation.

The reader does not need to grasp every detail in this chapter in order to proceed to subsequent ones. This is why the proofs of the propositions are relegated to the appendix. In fact, strictly speaking this chapter is not necessary for understanding subsequent ones. However, it is necessary to understand the basics in this chapter in order to achieve a coherent grasp of the properties of each model, regardless of how this model is used in the inference, and how the resulting algorithms are implemented.

1.2.3 Formulating the problem as optimal inference

In order to formulate the problem of recovering shape and reflectance from defocus and motion blur, we have to choose an image formation model among those derived in Chapter 2, and choose an inference criterion to be minimized. Typically, one describes the measured images via a model, which depends on the unknowns of interest and other factors that are judged to be important, and a criterion to measure the discrepancy between the measured image and the image generated by the model, which we call the "model image." Such a discrepancy is called "noise" even though it may not be related to actual "noise" in the measurements, but may lump all the effects of uncertainty and nuisance factors that are not explicitly modeled. We then seek to find the unknowns that minimize the "noise," that is, the discrepancy between the image and the model.

This view corresponds to using what statisticians call "generative models" in the sense that the model can be used to synthesize objects that live in the same space of the raw data. In other words, the "discrepancy measure" is computed at the level of the raw data. Alternatively, one could extract from the images various "statistics" (deterministic functions of the raw data), and exploit models that do not generate objects in the space of the images, but rather generate these statistics directly. In other words, the discrepancy between the model and the images does not occur at the level of the raw data, but, rather, at an intermediate representation which is obtained via computation on the raw data. This corresponds to what statisticians call "discriminative models;" typically the statistics computed on the raw data, also called "features," are chosen to minimize the effects of "nuisance factors" that are difficult to model explicitly, and that do not affect the outcome of the inference process. For instance, rather than comparing the raw intensity of the measured image and the model image, one could compare the output of a bank of filters that are insensitive, or even invariant, with respect to sensor gain and contrast.

1.2.4   Choice of optimization criteria, and the design of optimal algorithms

There are many different choices of discrepancy criteria that one can employ, and there is no "right" or "wrong" one. In particular, one could choose any number of function norms, or even functionals that are not norms, such as information-divergence. So, how does one choose one? This, in general, depends on the application. However, there are some general guidelines that one can follow for the design of generic algorithms, where the specifics of the problem are not known ahead of time.

For instance, one can go for simplicity, and choose the discrepancy measure that results in the simplest possible algorithms. This often results in the choice of a *least-squares criterion*. If simplicity is too humble a principle to satisfy the demanding reader, there are more sophisticated reasonings that motivate the use of least-squares criteria based on a number of axioms that any "reasonable" discrepancy measure should satisfy [Csiszár, 1991]. In Chapter 4 we show how to derive inference algorithms that are optimal in the sense of least squares.

Choosing a least-squares criterion is equivalent to assuming that the measured image is obtained from the model image by adding a Gaussian "noise" process. In other words, the uncertainty is additive and Gaussian. Although this choice is very common due to the simplicity of the algorithms it yields, it is conceptually incorrect, because Gaussian processes are not bounded nor positive, and therefore it is possible that a model image and a realization of the noise process may result in a measured image with negative intensity at certain pixels, which is obviously not physically plausible. Of course, if the standard deviation of the noise is small enough, the probability that the model will result in a negative image is negligible, but nevertheless the objection of principle stands.

Therefore, in Chapter 5 we derive algorithms that are based on a different noise model, which guarantees that all the quantities at play are positive. This is a Poisson noise, that is actually a more plausible model of the image formation process, where each pixel can be interpreted as a photon counter.

1.2.5   Variational approach to modeling and inference from accommodation cues

An observation is made early on, in Chapter 2, that the process of image formation via a finite aperture lens can be thought of as a diffusion process. One takes the "ideal image" (i.e., the radiance of the scene), properly warped to overlap with the image, and then "diffuses" it so as to obtain a properly blurred image to compare with the actual measurement. This observation is a mere curiosity in

Chapter 2. However, it has important implications, because it allows us to tap into the rich literature of variational methods in inverse problems. This results in a variety of algorithms, based on the numerical integration of partial differential equations, that are simple to implement and behave favorably in the presence of noise, although it is difficult to guarantee the performance by means of analysis.

The implementation of some of these algorithms is reported in the appendix, in the form of MATLAB® code. The reader can test first-hand what modeling choice and inference criteria best suits her scenario.

Motion blur also falls within the same modeling class, and can be dealt with in a unified manner, as we do in Chapters 7 and 8.

2

Basic models of image formation

The purpose of this chapter is to derive simple mathematical models of the image formation process. The reader who is not interested in the derivation can skip the details and go directly to equation (2.9). This is all the reader needs in order to proceed to Chapters 3 through 5. The reader may want to consult briefly Section 2.4 before Chapters 6 through 8.

2.1 The simplest imaging model

In this section we derive the simplest model of the image-formation process using an idealized lens.

2.1.1 The thin lens

Consider the schematic camera depicted in Figure 2.1. It is composed of a plane and a lens. The plane, called the *image plane*, contains an array of sensor elements that measure the amount of light incident on a particular location. The lens is a device that alters light propagation via diffraction. We assume it to be a "thin" (i.e., approximately planar) piece of transparent material that is symmetric about an axis, the *optical axis*, perpendicular to the image plane. The intersection of the optical axis with the plane of the lens is called the *optical center*. We call v the distance between the lens and the image plane. It can be shown [Born and Wolf, 1980] that energy emitted from an infinitesimal source at a distance u from the lens, going through the lens, converges onto the image plane only if u and v

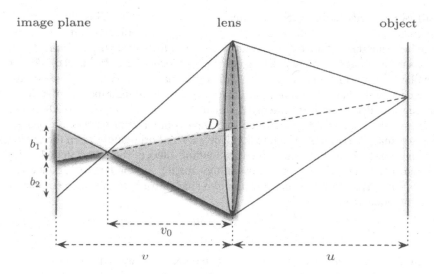

image plane   lens   object

Figure 2.1. The geometry of a finite-aperture imaging system. The camera is represented by a lens and an image plane. The lens has aperture D and focal length F. The thins lens conjugation law (2.1) is satisfied for parameters u and v_0. When $v \neq v_0$, a point in space generates an intensity pattern on the image plane. According to an idealized model based solely on geometric optics, the "point-spread function" is a disk of radius $b = b_1 = b_2 = D/2\,|1 - v/v_0|$; b is constant when the scene is made of an equifocal plane.

satisfy the *thin lens conjugation law*

$$\frac{1}{u} + \frac{1}{v} = \frac{1}{F}, \tag{2.1}$$

where F is a parameter that depends on the shape and material of the lens, called the *focal length*.[1] In this case we say that the source is *in focus*, and we call v, the distance between the lens and the image plane, the *focus setting*.

The thin lens law is a mathematical idealization; it can be derived from the geometry of a thin lens [Born and Wolf, 1980], or it can be deduced from the following axioms.

1. Any light ray passing through the center of the lens is undeflected.

2. Light rays emanating from a source at an infinite distance from the lens converge to a point.

---

[1]The focal length is computed via $1/F = (1/R_1 - 1/R_2)\,(n_2 - n_1)\,/n_1$ where n_1 and n_2 are the refractive indices of lens and air, respectively, and R_1 and R_2 are the radii of the two surfaces of the lens.

Call v_0 the focus setting such that equation (2.1) is satisfied. This simple model predicts that, if $v = v_0$, the image of a point is again a point. It is immediate to conclude that the surface made of points that are in focus is a plane, parallel to the lens and the image plane, which we call the *equifocal plane*. If, instead, the thin lens law is not satisfied (i.e., $v \neq v_0$), then the image of a point depends on the boundary of the lens. Assuming a circular lens, a point of light that is not in focus produces a disk of uniform intensity, called the *circle of confusion* (COC). The radius of the COC can be computed by simple similarity of triangles. Consider the two shaded triangles in Figure 2.1. The segment originating from the optical center ending on the plane lies on corresponding sides of these two triangles. By similarity of triangles, the ratio between the smaller and the larger sides is equal to $(v - v_0)/v_0$. Moreover, because the two shaded triangles are also similar by construction, we have

$$\frac{v - v_0}{v_0} = \frac{b_1}{D/2}, \tag{2.2}$$

where D is the *aperture* of the lens.[2] We can obtain the same equation for b_2, and conclude immediately that $b_1 = b_2 = b$. If $v_0 > v$ we only need to use $v_0 - v$ instead of $v - v_0$ in equation (2.2). This means that the COC is a disk and its radius b is given by

$$b = \frac{D}{2} \left| 1 - \frac{v}{v_0} \right|. \tag{2.3}$$

The indicator function of the COC normalized by its area is called the *pillbox function*.

It should be clear that the thin lens is a pure mathematical abstraction, because light propagation phenomena around finite apertures are subject to a variety of effects that we are not taking into account. The value of such an abstraction is to provide a model that is as simple as possible while supporting the task of inferring shape from images obtained with a real lens. In the next subsection we derive the simplest model of image formation that is the basis for us to develop inference algorithms to reconstruct 3-D shape.

2.1.2 Equifocal imaging model

The ideal image of a point light source, which we called the pillbox function, is a special case of the more general *point-spread function* that is a characteristic of the imaging system. Here we consider the simplest case, where the point-spread function is determined simply by the radius of the pillbox.

To make the notation more precise, notice that the pillbox h is a function defined on the image plane that takes values in the positive reals: $h : \Omega \subset \mathbb{R}^2 \rightarrow$

---

[2]The aperture D does not necessarily correspond to the diameter of the lens, because of the presence of a diaphragm, a mechanism that generates a variable-diameter opening in front of the lens. In standard cameras the aperture is determined by the F-number, or F-stop, F_0 via $D = F/F_0$.

$\mathbb{R}^+ \doteq [0, \infty)$ (Figure 2.1). Depending on the context, we may choose $\Omega \subset Z^2$ when we want to emphasize the discrete nature of the imaging sensor. In addition to its dependency on y, h also depends on the focus setting v of the optics, via the radius b in equation (2.3), and on the position of the point light source. We assume that the point source belongs to a surface s parameterized relative to the image plane by $s : \Omega \subset \mathbb{R}^2 \mapsto [0, \infty)$, so that we can write the 3-D position of a generic point source in the form $[x^T s(x)]^T$. To make the dependency of h from $\dot{x}$ and y explicit, we write $h^v(y, x)$. The superscript v is a reminder of the dependency on v, the focus setting. Suppose the point source in x emits light with infinitesimal intensity $r(x)dx$. Then the value of the pillbox function at the location y on the image plane is $h^v(y, x)r(x)dx$. The combination of the contributions from each point source forms the image. It is common to assume that sources combine linearly, so the image is simply the sum of the contribution from each source: $I(y) = \int h^v(y, x)r(x)dx$. Note that the image depends on $v \in \mathcal{V}$, the set of admissible focus settings. Therefore, $I : \Omega \times \mathcal{V} \mapsto [0, \infty); (y, v) \mapsto I^v(y)$ although occasionally we use the notation $I(y, v)$ or $I(y; v)$. To summarize, we have

$$I(y, v) = \int h^v(y, x)r(x)dx. \tag{2.4}$$

Here $h^v : \Omega \times \mathbb{R}^2 \mapsto [0, \infty)$ is called the *kernel* in equation (2.4), or the *point-spread function* (PSF) of the optics. When the scene is an equifocal plane, as in Figure 2.1, the radius b is constant and, therefore, the function h^v is *shift-invariant*; that is, it depends only on the difference $y - x$. With an abuse of notation we write

$$h^v(y - x) \doteq h^v(y, x). \tag{2.5}$$

The intensity $r(x)$ of a point source at x introduced in equation (2.4) is derived from a more complex quantity called *radiance*. A full derivation can be found in Appendix A. In the case of shape from defocus, it is not necessary to explicitly introduce radiance in the image formation model, because images are captured from the same point of view, and scene and illumination conditions are assumed static. Hence, we can approximate the radiance of the scene with a function $r :$ $\mathbb{R}^2 \mapsto [0, \infty)$ parameterized with respect to the image plane (for more details, refer to Appendix A). With a slight abuse of terminology, we call the function r radiance.

Remark 2.1. *A more accurate analysis of the image formation process should take diffraction into account. In this case, the image of a point in focus would not be a point, but a* diffraction pattern. *Optical instruments where diffraction effects are relevant are called* diffraction-limited *imaging systems, and are characterized by having high resolving power, such as microscopes and telescopes. The analysis of such systems requires the tools of Fourier optics [Goodman, 1996] which is beyond the scope of this book, where we limit our attention to geometric optics.*

In the study of shape from defocus, we typically capture a number K of images of the same scene with K different focus settings $[v_1 \ldots v_K]^T$, where each $v_i \in$

$\mathcal{V}$, $\forall\, i = 1 \dots K$. To simplify the notation, we collect all images $I(y, v_i)$ into a vector $\boldsymbol{I}(y) = [I(y, v_1) \dots I(y, v_K)]^T \in \mathbb{R}^K$, which we call the *multifocal vector image*. Similarly, we can collect the kernels h^{v_i} into the vector $\boldsymbol{h}(y, x) = [h^{v_1}(y, x) \dots h^{v_K}(y, x)]^T \in \mathbb{R}^K$. Then, equation (2.4) becomes

$$\boldsymbol{I}(y) = \int \boldsymbol{h}(y, x) r(x) dx. \tag{2.6}$$

However, for simplicity, we drop the boldface notation once we agree that $\boldsymbol{I}(y) \in \mathbb{R}^K$ and $\boldsymbol{h}(y, x) \in \mathbb{R}^K$, and call the multifocal vector image I and the vector kernel h simply I and h, respectively. In addition, when emphasizing the dependency on the shape s, we write

$$I_s(y) = \int h_s(y, x) r(x) dx. \tag{2.7}$$

When emphasizing the dependency on the radiance r, we write

$$I^r(y) = \int h(y, x) r(x) dx. \tag{2.8}$$

When there is no confusion, we simply use

$$\boxed{I(y) = \int_{\mathbb{R}^2} h(y, x) r(x) dx} \quad y \in \Omega \subset \mathbb{R}^2 \tag{2.9}$$

to denote the imaging model. In the equation above $y \in \Omega$ is a point on the image plane, $x \in \mathbb{R}^2$ and $h : \Omega \times \mathbb{R}^2 \mapsto \mathbb{R}^K$ is the kernel function or point-spread function; we (improperly) call the function $r : \mathbb{R}^2 \mapsto [0, \infty)$ radiance. The PSF also encodes the depth of the scene $s : \mathbb{R}^2 \mapsto [0, \infty)$ when there are no occlusions (see Section 2.3 for a suitable representation of the depth of the scene in the presence of occlusions). The *multifocal vector* image I generated by equation (2.9) is called *ideal image* or *noise-free image*.

2.1.3   Sensor noise and modeling errors

Images measured by a physical sensor are obviously different from the ideal image. In addition to sensor noise (most imaging sensors are photon counters and therefore are subject to shot noise), the discrepancy between the real and ideal images is due to all the unmodeled phenomena, from diffraction to lens aberrations, quantization, sampling errors, digitization, and so on. It is common to label the compound effect of all unmodeled phenomena as "noise" even though it is improper nomenclature. What is important is that the effect of unmodeled phenomena is described collectively with a probability distribution, rather than using a functional dependency. If we describe all unmodeled phenomena with a probability density p_n, then each measured image, or observed image, J, is the result of a particular realization of the noise process n; that is, it is a sample from a random process. The assumption we make is that all unmodeled phenomena tend to average out in the long term, so that the ideal image is the mean of the measured images:

$$I = \int J p_n(J) dJ. \qquad (2.10)$$

The simplest model of interaction of noise n with the measured image J is additive:

$$J = I + n. \qquad (2.11)$$

The characterization of the probability distribution p_n can be obtained by studying the physical properties of the sensor. Typically, for convenience, one assumes either a Gaussian or a Poisson model. In the rest of this book we explore both models, and derive inference algorithms that guarantee optimality with respect to the noise model above. In particular, the relationship between measured and ideal images in equation (2.10) allows us to treat Gaussian and Poisson noise in a unified manner.

2.1.4 Imaging models and linear operators

The basic model of image formation (2.9) may seem complex at first, especially for the reader not comfortable with integrals. In this section we show how one can think of (2.9) as a matrix equation, using properties of Hilbert spaces. This analogy is useful per se, and it is exploited in Chapter 4 to design reconstruction algorithms.

We start by assuming that the quantities r, h, I in (2.9) are square-integrable functions. This is not quite a germane assumption, and we challenge it in Chapter 3. However, for the moment let us indulge in it, and recall that the space of square-integrable functions, usually indicated by $L^2(\mathbb{R}^2) \doteq \{ f : \mathbb{R}^2 \to \mathbb{R} \mid \int f^2(x) dx < \infty \}$, is a Hilbert space with inner product $\langle\langle \cdot, \cdot \rangle\rangle : L^2 \times L^2 \longrightarrow \mathbb{R}$ defined by

$$(f, g) \mapsto \langle\langle f, g \rangle\rangle \doteq \int f(x) g(x) dx \qquad (2.12)$$

with the norm $\|f\| \doteq \sqrt{\langle\langle f, f \rangle\rangle}$. Consider also the case where Ω is a discrete domain, that is, $\Omega \equiv \mathbb{R}^{N \times M}$, where N is the number of columns of the CCD and M the number of rows. Because we collect K images by changing focus settings, the multifocal vector image $I \in \mathbb{R}^{K \times N \times M}$. Let $W = KMN$. Then, we introduce the ℓ^2 space $\mathbb{R}^W \sim \mathbb{R}^{K \times N \times M}$, with the inner product $\langle \cdot, \cdot \rangle : \mathbb{R}^W \times \mathbb{R}^W \longrightarrow \mathbb{R}$ defined by

$$(A, B) \mapsto \langle A, B \rangle \doteq \text{Trace}\{AB^T\} \qquad (2.13)$$

and norm $|A| = \sqrt{\langle A, A \rangle}$. If we interpret points in $\mathbb{R}^W$ as W-dimensional vectors, then the inner product is the usual $\langle a, b \rangle \doteq a^T b$ with $a, b \in \mathbb{R}^W$.

If we model the radiance r as a point in $L^2(\mathbb{R}^2)$, and the image I as a point in $\mathbb{R}^W$, then the imaging process, as introduced in equation (2.6), can be represented by an operator H_s,

$$H_s : L^2 \longrightarrow \mathbb{R}^W; \quad r \mapsto I = H_s r. \tag{2.14}$$

This equation is just another way of writing equation (2.6), one that is reminiscent of a linear system of equations, represented in matrix form. The following paragraph introduces operators that are useful in Chapter 4. The analogy with (finite-dimensional) systems of equations proves very useful in deriving algorithms to reconstruct shape from defocused images.

Adjoints and orthogonal projectors.

The operator $H_s : L^2 \to \mathbb{R}^W$ admits an adjoint H_s^* defined by the equation

$$\langle H_s r, I \rangle = \langle\langle r, H_s^* I \rangle\rangle \quad \forall \, r \in L^2 \tag{2.15}$$

from which we get that

$$H_s^* : \mathbb{R}^W \longrightarrow L^2; \quad I \mapsto \sum_{y \in \Omega} h_s(y, x) I(y). \tag{2.16}$$

Note that H_s^* is a function. The (Moore–Penrose) pseudo-inverse $H_s^\dagger : \mathbb{R}^W \longrightarrow L^2$ is defined such that $r = H_s^\dagger I$ solves the equation

$$H_s^* H_s r = H_s^* I \tag{2.17}$$

when it exists; with an abuse of notation we could write $H_s^\dagger = (H_s^* H_s)^{-1} H_s^*$. The orthogonal projector $H_s^\perp$ is then defined as

$$H_s^\perp : \mathbb{R}^W \longrightarrow \mathbb{R}^W; \quad I \mapsto H_s^\perp I = (I_d - H_s H_s^\dagger) I, \tag{2.18}$$

where I_d is the identity in $\mathbb{R}^{W \times W}$. Note that this is a finite-dimensional linear operator, represented therefore by a matrix.

2.2   Imaging occlusion-free objects

In the previous section we have seen that the image of a plane parallel to the lens is obtained via a linear shift-invariant model, equation (2.4). Unfortunately, the scene is seldom a plane and even less often equifocal. A more general scenario is when the scene is made of a continuous surface s that does not have self-occlusions. [3] The corresponding model, described in equation (2.9), is shift-varying.

In this section we reintroduce the imaging model from a different point of view. Rather than considering the image generated by a point in space, we consider the energy collected by a point on the image plane from light emitted by the scene. Figure 2.2 illustrates the geometry, which mirrors that introduced for the point-spread function and illustrated in Figure 2.1. We introduce this model because it

---

[3]Occlusions are modeled in the next section.

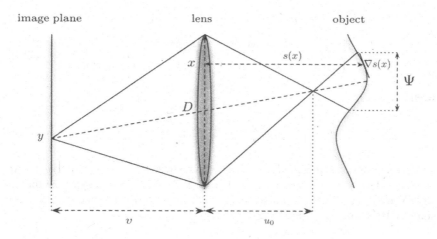

Figure 2.2. Geometry of occlusion-free imaging. In this configuration the thin lens law (2.1) is satisfied with parameters u_0 and v. The cone of light rays deflected by the lens to the point y on the image plane intersects the surface s on the region Ψ (see equation (2.20)). The sum of energy contributions from the cone determines the intensity measured at the point y. Compare with Figure 2.1.

is easy to extend it to deal with occlusions (Section 2.3), and because it has been shown to be accurate for real scenes [Asada et al., 1998b].

Objects in the scene can emit energy of their own, or transmit energy from third sources in ways that depend on their material. Rough wood or opaque matte surfaces, for instance, reflect light homogeneously in all directions (Lambertian materials), whereas polished stone or metal have a preferred direction. Even more complex is the case of *translucent materials* such as alabaster or skin, where light is partially absorbed and reflected after subsurface scattering. A more detailed description of materials and their reflective properties can be found in Appendix A.

In the context of shape from defocus we collect different images of the scene with different focus settings where illumination conditions and viewpoint remain unchanged. Therefore, we cannot distinguish between a self-luminous object and one that reflects light. Therefore, we only need to consider the amount of energy radiated from points in space, regardless of how it comes about. However, to keep the discussion simple, we assume that the surface s is Lambertian.[4] Under these assumptions, the imaging model for the geometry illustrated in Figure 2.2 is in the same form as in equation (2.4), but the kernel is no longer shift-invariant, and

---

[4]For our purposes it suffices that reflection is homogeneous along the cone of directions from the point to the lens, an assumption that is by and large satisfied by most materials.

is given instead by

$$h(y,x) = \begin{cases} \frac{u_0^2}{\pi(D/2)^2} \frac{|s(x)-u_0-\nabla s(x)^T\left(y\frac{u_0}{v}+x\right)|}{|s(x)-u_0|^3} & \forall\, x \in \Psi \\ 0 & \forall\, x \notin \Psi \end{cases}, \qquad (2.19)$$

where

$$\Psi = \left\{ x : \left\| y\frac{s(x)}{v} + x \right\| \le \frac{D}{2}\frac{|s(x)-u_0|}{u_0} \right\} \qquad (2.20)$$

and $\nabla s(x)$ denotes the gradient of s with respect to x, as we derive in Appendix A. For typical values of $u_0 \doteq 1/F - 1/v$ and D, it is easy to see that the term $\nabla s(x)^T\left(yu_0/v + x\right)$ can be neglected as long as the surface is sufficiently smooth. By doing this, we obtain

$$h(y,x) = \begin{cases} \frac{u_0^2}{\pi(D/2)^2} \frac{1}{|s(x)-u_0|^2} & \forall\, x \in \Psi \\ 0 & \forall\, x \notin \Psi. \end{cases} \qquad (2.21)$$

As a special case where $\dot{s}(x) = s$ (i.e., in the equifocal case), the kernel in equation (2.21) reduces to the pillbox kernel that we have discussed in the previous section.

2.2.1 Image formation nuisances and artifacts

Commercial cameras allow changing the focus setting by moving the lens relative to the sensor. Unfortunately, however, blurring is not the only effect of this procedure. The reader who is also an amateur photographer may have noticed that changing the focus also causes magnification artifacts. Such artifacts, unlike blur, do not contain any information about the 3-D structure of the scene and its radiance, so they are "nuisance factors" in the image formation process. Unfortunately we cannot just ignore them. The reader who tries to test the 3-D reconstruction algorithms described in the next chapters to images obtained simply by changing the focus setting in a commercial camera will be disappointed to see that they fail to work because the image formation model above is not accurate.

Fortunately, such artifacts are easy to eliminate by simple preprocessing or calibration procedures. In Appendix D, Section D.1, we discuss a simple calibration procedure that the reader can easily implement.

A more sophisticated approach would be to make magnification and other registration artifacts part of the unknowns that are estimated together with the 3-D shape and radiance of the scene. A yet more sophisticated approach, discussed in Appendix D, Section D.2, is to build a specialized imaging device that can generate defocused images without magnification artifacts. These devices are called *telecentric optical systems*.

Figure 2.3. The "pinhole prison:" Occluded portions of the background scene are not visible using a pinhole camera (left). Using a lens and a finite aperture, however, allows seeing through the bars (right).

2.3   Dealing with occlusions

The model described in previous sections, simplistic as it may be, is not the simplest one can use. In fact, if we let the aperture D become infinitesimal, light travels from a point on the scene through the "pinhole" aperture onto a unique point on the image plane. In this case we have simply that $I(y) = r(x)$, where x is a point on the ray through y and the origin, where the pinhole is located. This *pinhole imaging model* is clearly an idealization, because diffraction effects become dominant when the aperture decreases, and in the limit where the aperture becomes zero no light can pass through it. Nevertheless, it is a reasonable approximation for well-focused systems with large depth of field when the scene is at a distance much larger than the diameter of the lens (i.e., $s(x) \gg v$, $u_0 \; \forall \; x$), and it is the defacto standard in most computer vision applications. Where the pinhole imaging model breaks down is when the scene is at a distance comparable to the diameter of the lens. This phenomenon is dramatic at occlusions, as we illustrate in Figure 2.3. In both images, the bars on the windows occlude the sign on the mountain in the background. On the left, the picture has been taken with a pinhole camera (a camera with very small aperture D). On the right, the picture has been taken with a finite aperture and a lens focused on the background. Notice that, whereas in the "pinhole image" the occluded text is not visible and everything is in focus, in the "finite aperture image" one can read the text through the bars and the foreground is "blurred away." Clearly, the pinhole model can explain the first phenomenon, but not the second.

To illustrate imaging in the presence of occlusions, consider the setup in Figure 2.4. The intensity at a pixel $y \in \Omega$ on the image plane is obtained by integrating the contribution from portions of two objects. The size and shape of

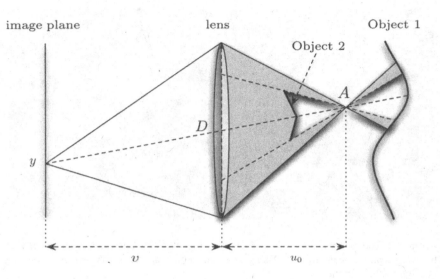

Figure 2.4. Imaging through an occlusion: The intensity measured on the image plane at the point y is the result of the light emitted from both Object 1 and Object 2. The radiance of Object 1 is integrated except for the region occluded by Object 2. This region depends on the projection of Object 1 onto Object 2 through the point in focus A.

these portions are determined by the mutual position of *Object 1* and *Object 2*, as well as on the geometry of the imaging system, as illustrated in Figure 2.4. In order to formalize the imaging model we need to introduce some notation. We denote with H the Heaviside function

$$H(z) = \begin{cases} 1, & \text{if } z \geq 0 \\ 0, & \text{if } z < 0 \end{cases} \tag{2.22}$$

and with $\bar{H}$ the complement of H; that is, $\bar{H} = 1 - H$ and

$$\bar{H}(z) = \begin{cases} 0, & \text{if } z \geq 0 \\ 1, & \text{if } z < 0. \end{cases} \tag{2.23}$$

The surface and radiance of Object 1 is denoted by the functions $s_1 : \mathbb{R}^2 \mapsto [0, \infty)$ and $r_1 : \mathbb{R}^2 \mapsto [0, \infty)$. Similarly, $s_2 : \mathbb{R}^2 \mapsto [0, \infty)$ and $r_2 : \mathbb{R}^2 \mapsto [0, \infty)$ correspond to Object 2. Let ϖ be the projection of s_1 on s_2 through the point A of coordinates $(-yu_0/v, u_0)$ (see Figure 2.4). Object 2 does not occupy the entire space $\mathbb{R}^2$, but only a subset $\Omega_2 \subset \mathbb{R}^2$. Notice that Ω_2 could be a simple connected set, or a collection of several connected components. To represent such diverse topologies, and to be able to switch from one to another in a continuous fashion, we define Ω_2 implicitly as the zero level set of a function $\phi : \mathbb{R}^2 \mapsto \mathbb{R}$, that we call the *support function*. ϕ defines the domain Ω_2 of r_2 and s_2 via

$$\Omega_2 = \{ x \in \mathbb{R}^2 : \phi(x) \geq 0 \} \tag{2.24}$$

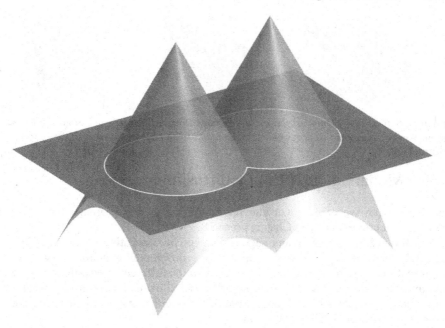

Figure 2.5. Representing the support function ϕ: The domain Ω_2 is defined as the zero level set of ϕ and is shown as a "figure eight."

and it is a *signed distance function*; that is, it satisfies $|\nabla\phi(x)| = 1$ almost everywhere (a.e.) [Sethian, 1996]. An example of Ω_2 is shown in Figure 2.5. Notice how one can make Ω_2, which is a single-connected component in Figure 2.5, a set composed of two connected components by shifting the support ϕ down. Using the notation introduced we can write equation (2.24) as

$$\Omega_2 = \{x \in \mathbb{R}^2 : H\big(\phi(x)\big) = 1\}. \tag{2.25}$$

We are now ready to introduce the imaging model for occlusions. From the image formation model (2.19) introduced in the previous section, we have that the image $I : \Omega \mapsto [0, \infty)$ can be expressed as

$$I(y) \quad = \quad \int_{\Omega_2} h_{s_2}(y, x) r_2(x) dx \\ + \int_{\mathbb{R}^2} h_{s_1}(y, x) \bar{H}\big(\phi(\varpi(x))\big) r_1(x) dx \tag{2.26}$$

or, equivalently, by

$$\boxed{I(y) \quad = \quad \int_{\mathbb{R}^2} h_{s_2}(y, x) H\big(\phi(x)\big) r_2(x) dx \\ + \int_{\mathbb{R}^2} h_{s_1}(y, x) \bar{H}\big(\phi(\varpi(x))\big) r_1(x) dx.} \tag{2.27}$$

Figure 2.6. Images as heat distributions, blurred by diffusion. The radiance of an equifocal scene can be interpreted as heat on a metal plate. Diffusion causes blurring of the original distribution.

2.4   Modeling defocus as a diffusion process

In this section we explore yet another interpretation of the image formation process that comes handy in Chapters 6 and 8.

Equation (2.9) shows that an image is obtained by computing an integral. This integral can be thought of as the solution of a differential equation, so we can represent the imaging process equivalently with the differential equation, or with the integral that solves it. This interpretation will allow us to tap into the wealth of numerical schemes for solving partial differential equations (PDEs) , and we now explore it using a physical analogy.

The image in (2.9) is just a blurred version of the radiance of the scene. If we think of the radiance as describing the heat distribution on a metal plate, with the temperature representing the intensity at a given point, then heat diffusion on the plate causes blurring of the original heat distribution (Figure 2.6). Thus an image can be thought of as a diffusion of the radiance according to heat propagation; the more "time" goes by, the blurrier the image is, until the plate has uniform temperature. This process can be simulated by solving the *heat equation*, a particular kind of partial differential equation, with "time" being an index that describes the amount of blur. In practice, blurring – or from now on "diffusion" – is not the same across the entire image because the scene is not flat. The amount of diffusion depends on the geometry of the scene through a space-varying diffusion coefficient .

Before we further elaborate on this analogy, which we do in the next subsections, we need to discuss our choice of the pillbox as a point spread function. This is only a very coarse approximation for real imaging systems, that is valid only for very large apertures. In practice, real point-spread functions for the type of commercial cameras that we normally use are better approximated by a smoothed version of a pillbox (Figure 2.7 top), that becomes close to a Gaussian function for small apertures (Figure 2.7 bottom). A Gaussian function, or more precisely the circularly symmetric 2-D Gaussian shift-invariant kernel

$$h(y, x) = \frac{1}{2\pi\sigma^2} e^{-\|x-y\|^2/(2\sigma^2)} \tag{2.28}$$

is a common model for the PSF of most commercial cameras. It has been argued that this model is valid even when considering diffraction effects (see Remark 2.1

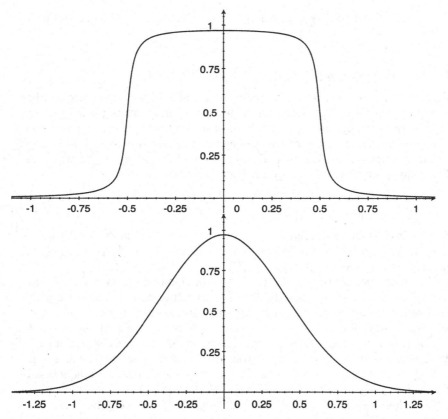

Figure 2.7. Real point-spread functions. One can get an approximate measurement of a PSF by taking an image of a small distant light source. For large apertures (top) the PSF is approximated by a smoothed version of a pillbox, but when the aperture becomes smaller (bottom) the PSF is better approximated by a Gaussian.

and [Pentland, 1987], [Subbarao, 1988], and [Chaudhuri and Rajagopalan, 1999]). One can even modify the optical system by placing suitable photographic masks in front of the lens to make the PSF Gaussian. More pragmatically, the Gaussian is a very convenient model that makes our analysis simple. In the equation above, the *blurring parameter* $\sigma = \gamma b$, γ is a calibration parameter for the designer to determine and $b = D/2\,|1 - v/v_0|$ is the blurring radius.

More important than the actual shape of the PSF is the fact that it has to satisfy the *energy conservation principle* (or normalization):

$$\int_{\Omega} h_s(y, x)dy = 1 \qquad \forall\, x \in \mathbb{R}^2 \tag{2.29}$$

for any surface s and focus-settings. This is equivalent to requiring that our optical system be lossless, so that all the energy emitted by a point in the scene is transferred to the image plane and our camera does not absorb or dissipate energy.

We now resume our physical analogy and consider the case of a flat (equifocal) scene.

2.4.1 Equifocal imaging as isotropic diffusion

Suppose our scene is a flat metal plate, with a heat distribution corresponding to its radiance; that is, the temperature at a point represents the radiant energy. Then blurring can be simulated by heat diffusing on the plate, which is governed, assuming the plate to be homogeneous, by the so-called *isotropic heat equation*. [5] Starting from the radiance r as initial condition, heat propagates in such a way that at any given time t, the temperature at a position x satisfies:

$$\begin{cases} \dot{u}(x,t) = c\triangle u(x,t) & c \in [0,\infty) \quad t \in [0,\infty) \\ u(x,0) = r(x) & \forall\, x \in \mathbb{R}^2. \end{cases} \tag{2.30}$$

Here the dot denotes differentiation in time, $\dot{u} \doteq \partial u / \partial t$, and the symbol $\triangle$ denotes the Laplacian operator $\sum_{i=1}^{2} \partial^2 / \partial x_i^2$ with $x \doteq [x_1\ x_2]^T$. The nonnegative parameter c is called the *diffusion coefficient*.

The reader can verify by substitution that the image in (2.4) solves this partial differential equation when the PSF is the shift-invariant Gaussian function of equation (2.28). More precisely, the image captured with a particular focus setting v is the solution $u : \mathbb{R}^2 \times [0,\infty) \mapsto [0,\infty)$ at a specific time $t = t_1$, $I(y,v) = u(y,t_1), \forall\, y \in \Omega$, where v is related to the diffusion coefficient c (and t_1) as we soon show. Because we restrict our attention to solutions that are spatially bounded, such as images, we are guaranteed existence and uniqueness of the solution to equation (2.30) [Weickert et al., 1998].

The solution u can also be represented via *Green's function*. The Green function $G : \Omega \times \mathbb{R}^2 \times [0,\infty) \mapsto [0,\infty)$ corresponding to equation (2.30) is defined as the impulse response of equation (2.30); that is, $G(y,x,t) \doteq u(y,t)$ where u is the solution of equation (2.30) with initial conditions $u(y,0) = \delta(x-y)$, and $\delta(\cdot)$ is Dirac's delta. The function G satisfies the equation

$$u(y,t) = \int G(y,x,t)u(x,0)dx = \int G(y,x,t)r(x)dx. \tag{2.31}$$

The Green function relative to the isotropic heat equation (2.30) is a Gaussian with the blurring parameter σ related to the diffusion coefficient via

$$\sigma^2 = 2tc, \tag{2.32}$$

as can be verified by direct substitution of equation (2.31) into equation (2.30). This explains why we so cherish Gaussians as point-spread functions, beyond the discussion leading to this section. Gaussians establish the equivalence between

---

[5]Notational *mea culpa:* we use u for the solution of the isotropic heat equation as is customary in the PDE literature. Alas, we have already used u to describe the geometry of the lens. Fortunately, there should be little confusion because the solution of equation (2.30) is a function of space and time, whereas the optical setting u is a scalar value.

the integral imaging model (2.9) and its differential counterpart (2.30) via (2.32). Notice that the diffusion coefficient c and the time variable t are determined up to a common factor, which is easily resolved by fixing a nominal time, say $t = 1/2$.

Now, what about scenes that are not equifocal? Does the analogy of a homogeneous metal plate still hold?

2.4.2 Nonequifocal imaging model

We have seen in Section 2.2 that, when the surface s is not an equifocal plane, the corresponding PSF is in general shift-varying, so the interpretation of the imaging process as the solution of the heat equation does not hold. However, the analogy can still help us in that, rather than seeking an approximation to the shift-varying PSF, we can generalize the isotropic heat equation to one that satisfies the energy conservation principle (2.29) while being space-varying. This generalization yields the *inhomogeneous diffusion equation* [6] and uses a space-varying diffusion coefficient[7] $c \in C^1(\mathbb{R}^2)$, and $c(x) \geq 0 \; \forall \, x \in \mathbb{R}^2$. Furthermore, in order to satisfy the energy conservation requirement in Proposition 2.2, we need to assume that the subset $\{x : c(x) > 0\}$ of the image domain Ω, is in C^1 (i.e. it is bounded and its boundary is locally mapped by functions in $C^1(\mathbb{R})$). The inhomogeneous diffusion equation is then defined as

$$\begin{cases} \dot{u}(x,t) = \nabla \cdot (c(x)\nabla u(x,t)) & t \in [0,\infty) \\ u(x,0) = r(x) & \forall \, x \in \mathbb{R}^2 \end{cases} . \qquad (2.34)$$

where the symbol ∇ is the gradient operator $[\partial/\partial x_1 \;\; \partial/\partial x_2]^T$ with $x = [x_1 \; x_2]^T$, and the symbol $(\nabla \cdot)$ is the divergence operator $\sum_{i=1}^2 \partial/\partial x_i$. The next proposition shows that equation (2.34) satisfies the energy conservation principle (2.29). To see it, we need to use the divergence theorem (see [Gilardi, 1992]), and our assumptions on the diffusion coefficient. It also helps to notice that the energy conservation requirement can be rephrased as follows.

$$\int u(y,t)dy = \iint G(y,x,t)u(x,0)dxdy = \int u(x,0)dx \qquad (2.35)$$

for any nonnegative function $u(x,0)$. In other words, the spatial average of u does not change in time. We can now state and prove energy conservation.

---

[6]Note that the *inhomogeneous diffusion equation* is different from what in some literature is called *nonhomogeneous diffusion equation*, which is characterized by a forcing term added to the classical heat equation, for instance,

$$\begin{cases} \dot{u}(x,t) = c\triangle u(x,t) + f(x,t) & t \in [0,\infty) \\ u(x,0) = r(x) & \forall \, x \in \mathbb{R}^2 \end{cases} , \qquad (2.33)$$

where $f(x,t)$ is the forcing term. The nomenclature *inhomogeneous* emphasizes that the diffusion coefficient is not homogeneous in space.

[7]$C^1(\mathbb{R}^2)$ is the space of scalar functions with continuous partial derivatives in $\mathbb{R}^2$.

Proposition 2.2. *The solution of the inhomogeneous heat equation* (2.34) *with* $c : \mathbb{R}^2 \mapsto [0, \infty), c \in C^1$ *satisfies the energy conservation principle if the open set* $\mathcal{O} = \{x : c(x) > 0\}$ *is such that* $\mathcal{O} \in C^1$.

Proof. It suffices to show that the integral of the solution of the heat equation above does not change in time. To this end we write:

$$
\int \dot{u}(x,t)dx = \int \nabla \cdot (c(x)\nabla u(x,t))dx = \int_{\mathcal{O}} \nabla \cdot (c(x)\nabla u(x,t))dx
$$
$$
= \int_{\partial\mathcal{O}} c(x)\nabla u(x,t) \cdot n \, dl = 0,
$$

$$(2.36)$$

where $\partial\mathcal{O}$ denotes the boundary of $\mathcal{O}$, n denotes the outer normal to $\partial\mathcal{O}$, and dl is the line element along the boundary. The first equality comes from the model in equation (2.34). The second step is due to the fact that $c(x) = 0$ for $x \notin \mathcal{O}$. The third step is a consequence of the divergence theorem and the assumption on $\mathcal{O}$ and $\partial\mathcal{O}$. Finally, the last equality comes from the fact that $c(x) = 0$ for $x \in \partial\mathcal{O}$. □

By assuming that the surface s is smooth, we can relate again the diffusion coefficient c to the space-varying blurring coefficient σ via

$$
\sigma^2(x) \simeq 2tc(x). \tag{2.37}
$$

Therefore, we can represent the image of a nonequifocal scene as the solution of a partial differential equation, where the diffusion coefficient c is related to the focus setting. This interpretation comes in handy in Chapters 6 and 8, where we tap efficient numerical schemes to solve partial differential equations in order to implement shape reconstruction algorithms.

2.5 Modeling motion blur

The value of a blurred image at a particular point (pixel) is obtained by averaging the radiance over a region of space determined by the optics and shape of the scene. In practice, the image is obtained by averaging the photon count over time during the shutter interval . If the scene is moving relative to the camera, temporal averaging results in blurring even if the scene is in focus. This effect is commonly known as *motion blur*. In this section we derive a model of image formation that takes into account motion blur, and in the next section we show how to model spatial and temporal averaging simultaneously. In particular, in Section 2.5.2 we show that the interpretation of blurring as a diffusion allows for a particularly simple unification of the two effects.

2.5.1 *Motion blur as temporal averaging*

Suppose we capture an image of a scene populated by a number of objects that are moving independently. If their motion is faster than the shutter speed of the

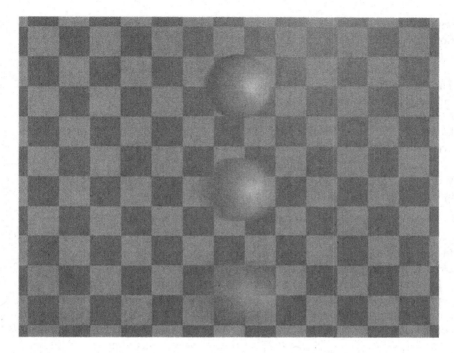

Figure 2.8. Motion blur. Objects moving faster than the shutter appear motion blurred. The amount of motion blur depends on the shutter speed as well as the objects speed.

camera, then they will appear *motion blurred* (Figure 2.8). To derive a first simple model of the formation of a motion-blurred image, assume that the scene is made of a single rigid object, and the camera moves relative to it. This generates a relative image motion which we call ν and that we characterize later in the section. For now, $\nu : \Omega \times [0, \infty) \to \mathbb{R}^2$ is a two-dimensional function (field) of the pixel coordinate y and time t. If the shutter remains open for a time interval[8] ΔT, then the image I is obtained by averaging the radiance over ΔT:

$$I(y) = \frac{1}{\Delta T} \int_{-\Delta T/2}^{\Delta T/2} r(y + \nu(y, t)) dt. \qquad (2.38)$$

Note that we do not consider saturation and we assume that images are normalized with respect to the shutter interval, so as to have the same average energy regardless of shutter speed. For small amounts of blur (slow motion or fast shutter), we can approximate the model above with the following

$$I(y) \simeq \int \frac{1}{\sqrt{2\pi}\Delta\tilde{T}} e^{-t^2/(2\Delta\tilde{T}^2)} r(y + \nu(y, t)) dt \qquad (2.39)$$

---

[8]The symbol Δ represents the *difference operator*.

with $\Delta \tilde{T} = \kappa \Delta T$ and κ is a suitable constant that can be determined by a calibration procedure.[9] This is equivalent to averaging over a "soft temporal window" rather than in between the idealized opening and closing of the shutter, because in reality the shutter, under the auspices of inertia, opens and closes gradually. The constant κ may be necessary to match $\Delta \tilde{T}$ with the nominal shutter interval indicated in the camera specifications.

Now consider a scene composed of Y objects moving independently with velocities $\nu_1, \ldots, \nu_Y$. Denote with $\{\Omega_j\}_{j=1\ldots Y}$ the regions of the image plane occupied by the projections of each of the moving objects. Note that $\{\Omega_j\}_{j=1\ldots Y}$ is a partition of Ω, that is, that $\Omega = \bigcup_{j=1}^{Y} \Omega_j$ and that $\Omega_j \bigcap \Omega_i = \emptyset$ for $\forall\, i, j = 1 \ldots Y, i \neq j$. This is because the "background" scene can be considered to be an object itself. In this case, the image model becomes

$$I(y) = \int \frac{1}{\sqrt{2\pi}\Delta\tilde{T}} e^{-t^2/(2\Delta\tilde{T}^2)} r(y + \nu_j(y,t)) dt \qquad \forall\, y \in \Omega_j. \qquad (2.40)$$

In order to complete the model we need to determine the velocities ν_j, which are a function of the shape and motion of objects in the scene. To do so consider again, for simplicity, the case of a scene made of a single object with depth map s. If the object is rigid, its spatial velocity can be characterized by a translational component $V(t) = [V_1(t)\ V_2(t)\ V_3(t)]^T \in \mathbb{R}^3$ and a rotational component $\omega(t) = [\omega_1(t)\ \omega_2(t)\ \omega_3(t)]^T \in \mathbb{R}^3$, so that a point with coordinates $X(t) = [X_1(t)\ X_2(t)\ X_3(t)] \in \mathbb{R}^3$ at time t evolves according to a rigid body motion:

$$\dot{X}(t) = \omega(t) \times X(t) + V(t), \qquad (2.41)$$

where $\times$ denotes the cross-product [10] of two vectors in $\mathbb{R}^3$. Note that points $X(t)$ can move relative to the camera because the scene is moving rigidly, or because the scene is static and the camera is moving relative to it. All that matters is the relative (rigid) motion between the two (see [Ma et al., 2003], Chapter 3 for more details).

The velocity of the projection of points in space onto the image plane can be computed directly by taking the temporal derivative of the canonical projection $x(t)$:

$$x(t) \doteq [x_1(t)\ x_2(t)]^T \doteq v \left[\frac{X_1(t)}{X_3(t)}\ \frac{X_2(t)}{X_3(t)} \right]^T \qquad (2.42)$$

and substituting the rigid body motion into it. Here v is the focus setting relative to the first input image. It is the distance of the image plane from the lens for

---

[9]For instance, fitting a pillbox with a Gaussian using the L^1 norm yields $\kappa = 0.336$.

[10]Given two vectors $u, v \in \mathbb{R}^3$, their *cross-product* is a third vector with coordinates given by

$$u \times v \doteq \begin{bmatrix} u_2 v_3 - u_3 v_2 \\ u_3 v_1 - u_1 v_3 \\ u_1 v_2 - u_2 v_1 \end{bmatrix} \in \mathbb{R}^3.$$

the first defocused image we collect (see Remark 2.3 on how to collect motion-blurred images.) Note that, in the case of a pinhole model, v is substituted by the focal length F. To simplify the derivation one can write the coordinates of a point $X(t)$ as a function of their projection $x(t)$,

$$X(t) = \left[\frac{x(t)^T}{v} \; 1 \right]^T s(x(t)). \tag{2.43}$$

Then, by direct substitution, one obtains the velocity at a pixel x at time t:

$$
\dot{x}(t) = \frac{1}{s(x(t))} \begin{bmatrix} v & 0 & -x_1(t) \\ 0 & v & -x_2(t) \end{bmatrix} V(t) \\
+ \begin{bmatrix} -\frac{x_1(t)x_2(t)}{v} & v + \frac{x_1^2(t)}{v} & -x_2(t) \\ -\left(v - \frac{x_2^2(t)}{v}\right) & \frac{x_1(t)x_2(t)}{v} & x_1(t) \end{bmatrix} \omega(t). \tag{2.44}
$$

If instead of following a point as it moves on the image we stand still at a fixed pixel $x \in \Omega$, and look at the velocity of particles passing through it, we obtain the *velocity field* (or motion field) $\nu(x, t) \doteq \dot{x}(t)$, still a function of time, as

$$
\nu(x, t) \doteq \frac{1}{s(x)} \begin{bmatrix} v & 0 & -x_1 \\ 0 & v & -x_2 \end{bmatrix} V(t) \\
+ \begin{bmatrix} -\frac{x_1 x_2}{v} & v + \frac{x_1^2}{v} & -x_2 \\ -\left(v - \frac{x_2^2}{v}\right) & \frac{x_1 x_2}{v} & x_1 \end{bmatrix} \omega(t). \tag{2.45}
$$

For simplicity, in the rest of the book we restrict our attention to a much simplified model where the scene only undergoes fronto-parallel translation locally in space and time, so that we only have two degrees of freedom rather than six:

$$\nu(x, t) = v\frac{V_{1,2}(t)}{s(x)}, \tag{2.46}$$

where $V_{1,2}$ is the vector made of the first two components of the velocity V. Then the image velocity $\nu_j(x, t)$ of each moving object in the scene is given by

$$\nu_j(x, t) = v\frac{V_{1,2}^j(t)}{s(x)} \qquad \forall \, x \in \Omega_j, \tag{2.47}$$

where $V_{1,2}^j$ is the spatial velocity of the jth object; occasionally we omit the argument x in $\nu_j(x, t)$ for simplicity. When velocity $V(t)$ is constant, we also omit the argument t. Notice that there is a scale ambiguity in determining the velocity $V_{1,2}^j$ and the depth map s given the image velocity ν. This ambiguity can be resolved by assigning, for instance, the mean depth to a prescribed value. Finally, we arrive at the following model for the intensity of a motion-blurred image at a fixed pixel $y \in \Omega_j$.

$$I(y) = \int \frac{1}{\sqrt{2\pi}} e^{-t^2/2} r\left(y + \Delta\tilde{T}v\frac{V_{1,2}^j}{s(y)}t\right) dt \qquad \forall \, y \in \Omega_j. \tag{2.48}$$

Remark 2.3 (On capturing motion-blurred images). *How can we capture motion-blurred images? Some consumer cameras come equipped with so-called auto-bracketing, that is, the capability of automatically capturing multiple images of the same scene with different exposures during the same session. However, even a cheaper automatic camera can be used, by following a simple procedure. First collect a sequence of images. Time averaging the sequence simulates a long shutter interval. For example, one could collect three images $[\bar{J}_1, \bar{J}_2, \bar{J}_3]$, and then consider $J_1 = \bar{J}_2$ as one motion-blurred image and $J_2 = 1/3 \sum_{i=1}^{3} \bar{J}_i$ as the second motion-blurred image. The shutter interval for the second image J_2 is approximately three times the shutter interval of the first image J_1. In this case, the data collection is rather simple, because no alignment and no synchronization are required, but it is based on the assumption that motion does not change among the three frames.*

2.5.2 Modeling defocus and motion blur simultaneously

In this section, we consider images where defocus and motion blur occur simultaneously. In the presence of motion, a defocused image $\hat{I}$ measured at time t can be expressed as

$$\hat{I}(y + \nu_j(y,t)) = \int h(y + \nu_j(y,t), x) r(x) dx \qquad \forall\, y \in \Omega_j. \tag{2.49}$$

Now, if we approximate both defocus and motion blur with a Gaussian kernel as in equation (2.39), then we obtain

$$
\begin{aligned}
I(y) &= \int \frac{1}{\sqrt{2\pi}} e^{-t^2/2} \hat{I}(y + \nu_j(y,t)) dt \\
&= \int \frac{1}{\sqrt{2\pi}} e^{-t^2/2} \int \frac{1}{2\pi\sigma^2} e^{-\|y - x + \Delta\tilde{T}\nu_j(y,t)\|^2/(2\sigma^2)} r(x) dx dt
\end{aligned}
\tag{2.50}
$$

for all $y \in \Omega_j$. If we now change the order of integration, we can write the previous equation in a more compact form as

$$I(y) = \int \frac{1}{2\pi|\Sigma|} e^{-(y-x)^T \Sigma^{-1}(y-x)/2} r(x) dx \tag{2.51}$$

where the variance tensor $\Sigma = \sigma^2 I_d + \nu_j \nu_j^T \Delta\tilde{T}^2$, I_d denotes the 2×2 identity matrix, and $|\Sigma|$ is the determinant of Σ. Equation (2.51) can also be seen as the solution of the following anisotropic diffusion equation:

$$
\begin{cases}
\dot{u}(x,t) = \nabla \cdot (D \nabla u(x,t)) & t \in (0, \infty) \\
u(x,0) = r(x)
\end{cases}
\tag{2.52}
$$

with the radiance r as initial condition and diffusion tensor

$$D = \frac{1}{4t}\Sigma. \tag{2.53}$$

In particular, if we consider the solution at time $t = 1/4$, then $D = \Sigma = \sigma^2 I_d + \nu_j \nu_j^T \Delta\tilde{T}^2$. Now, we can extend the model to the space-varying case, so that locally

$$D(x) = \sigma^2(x)I_d + \nu_j(x,t)\nu_j(x,t)^T \Delta \tilde{T}^2 \qquad (2.54)$$

and equation (2.52) becomes the following inhomogeneous anisotropic diffusion equation:

$$\begin{cases} \dot{u}(x,t) = \nabla \cdot (D(x)\nabla u(x,t)) & t \in (0,\infty) \\ u(x,0) = r(x). \end{cases} \qquad (2.55)$$

By substituting equation (2.46) into equation (2.54) and by including the shutter time interval $\Delta \tilde{T}$ in the velocity vector V, we have

$$D(x) = \sigma^2(x)I_d + v^2 \Delta \tilde{T}^2 \frac{V_{1,2}^j \left(V_{1,2}^j\right)^T}{s^2(x)}. \qquad (2.56)$$

Notice that the diffusion tensor just defined is made of two terms: $\sigma^2(x)I_d$ and $V_{1,2}^j(V_{1,2}^j)^T/s^2(x)$. The first term corresponds to the isotropic component of the tensor, and captures defocus. The second term corresponds to the anisotropic component of the tensor, and captures motion blur. Furthermore, because both terms are guaranteed to be always positive-semidefinite, [11] the diffusion tensor in equation (2.56) is positive semidefinite as well.

2.6 Summary

In this chapter we have presented simple models of the imaging process that we use in deriving shape from defocus algorithms. The value of the image at a particular location is obtained by integrating the energy coming from a region of space that depends upon the scene geometry and the imaging device. The shape and size of the region determine a kernel, that modulates the radiance of the scene. For scenes that are planar and parallel to the image plane, the kernel is shift-invariant; that is, the region of space over which the integral is performed does not change shape or size. When the scene is a general surface, without self-occlusions, the kernel is shift-varying. In the presence of occlusions, the model can be generalized using indicator functions, yielding a discontinuous imaging model.

The integral that yields an image can also be thought of as the solution of a partial differential equation. For the case of a Gaussian point-spread function, such a PDE is the classical heat equation for planar equifocal scenes, and can be generalized to nonequifocal scenes.

All of these models describe the "direct" process of how images are formed for a given scene and camera geometry. In the next chapters we instead face the inverse problem: given images we want to infer the unknowns of the scene. This

---

[11] A square matrix $A \in \mathbb{R}^{N \times N}$ is *positive definite* if $\forall v \in \mathbb{R}^N$ we have $v^T A v > 0$. A is *positive semidefinite* if $\forall v \in \mathbb{R}^N$ we have $v^T A v \geq 0$.

task is difficult due to the loss of information caused by the image formation process. Hence, one may wonder whether the inverse problem can be solved at all. In the next chapter we address precisely this question.

Before the reader can successfully reconstruct 3-D structure and radiance using the algorithms described in the next few chapters, it is necessary to precalibrate the images in order to eliminate magnification artifacts and other errors in the registration of different blurred images. This issue is addressed in more detail in Appendix D.

3

Some analysis: When can 3-D shape be reconstructed from blurred images?

In the previous chapter we have seen how a scene with a certain shape, represented by a function s, and a certain radiance r, can generate an image I under a certain focus setting, summarized in equation (2.4). Our main concern from now on is to use this equation to try to infer shape and radiance given a number of images taken with different settings. Before we venture into the design of algorithms to infer shape and radiance from blurred images, however, we need to ask ourselves whether this problem can be solved at all. It is easy to concoct scenes that cannot be reconstructed based solely on defocus information. Take, for instance, a white cube illuminated uniformly on a white background. We will never be able to tell it apart from a white sphere on a white background, no matter how many defocused images we take. So, clearly, whether we can reconstruct 3-D shape depends on the radiance of the scene, which we do not know. Under what conditions, then, can we reconstruct shape and radiance? In this chapter we address this question and its many facets. We organize our analysis around the role of radiance. We first assume that it can be chosen purposefully (Sections 3.2 and 3.3), and later relax it to be arbitrary (Section 3.4). Along the way, we point out some issues concerning the hypothesis on the radiance in the design of algorithms for reconstructing shape from defocus (Section 3.3.1).

In addition to the general ambiguities that we describe in this chapter, even in the idealized noise-free case there is no unique solution of the problem of reconstructing shape and radiance from defocused images because the unknowns are infinite-dimensional, whereas the images are digitized and quantized to finite dimensions. In fact, shape from defocus is a particular case of *blind deconvolution*, that is an ill-posed problem, as we discuss in Appendix F. This aspect of the problem is addressed via regularization, which we discuss in Section 3.5.

An important issue that is described in this chapter is the choice of objective function to be minimized in the reconstruction of shape and radiance. In Section 3.6 we discuss the two most common discrepancy measures that are used in the sequel of the book.

The reader interested in the reconstruction algorithms can skip this chapter at a first reading, except for Section 3.6 and the summary at the end of the chapter. However, eventually the user of shape from defocus algorithms will need to understand and appreciate the analysis in this chapter in order to make sense of the results of the algorithms. Where the conditions for reconstruction are not satisfied, any algorithm will return nonsense. For simplicity, in this chapter we restrict our attention to the case of defocus. Some of the basic ideas apply also for the case of motion blur, but we defer the analysis of that case to Chapters 7 and 8.

3.1   The problem of shape from defocus

Recall the imaging model[1] given in equation (2.4):

$$I_s^r(y, v) = \int h_s^v(y, x) r(x) dx, \tag{3.1}$$

where $x, y \in \mathbb{R}^2$, $r : \mathbb{R}^2 \mapsto [0, \infty)$ represents the radiance of the scene (see Appendix A, Section A.1.2), $s : \mathbb{R}^2 \mapsto [0, \infty)$ is a smooth function representing the depth map of the scene (i.e., the scene is a parameterized surface $\{(x, s(x)), \ x \in \Omega \subset \mathbb{R}^2\}$), h_s^v is the kernel, or point-spread function, and v is a focus setting belonging to a suitable set $\mathcal{V}$ (see Chapter 2, Section 2.1). Given the radiance r and the depth map s, which represents the "shape" or "surface" of the scene, we can use equation (3.1) to *generate* a defocused image I_s^r. This is the *direct problem*. In this book, we are interested in the *inverse problem* instead. Given a collection of K defocused images $I_s^r(\cdot, v_1) \ldots I_s^r(\cdot, v_K)$ corresponding to different focus settings $v_1 \ldots v_K$, we want to infer the shape s of the scene that generated the images $I_s^r(\cdot, v_i)$, $i = 1 \ldots K$. In this context, the radiance r is a *nuisance factor*,[2] that is, an unknown quantity that we are not directly interested in, but which affects the measurements. This problem is called *shape from defocus*. Occasionally we may be interested in inferring the radiance as well, in which case we talk about *deblurring*, or *image restoration* .

As mentioned in Chapter 2, the kernel h is a function $h_s^v : \Omega \times \mathbb{R}^2 \mapsto [0, \infty)$ where $\Omega \subset \mathbb{R}^2$ is a bounded subset of $\mathbb{R}^2$. For instance, we may consider kernels belonging to the Gaussian family

$$h_s^v(y, x) = \frac{1}{2\pi\sigma^2} e^{-\|x-y\|^2/(2\sigma^2)}. \tag{3.2}$$

---

[1] Note that the variables x and y are two-dimensional vectors of real numbers.

[2] In some instances, in order to infer the surface s we may have to retrieve r as well.

where σ is a function of both v and s, or the pillbox function introduced in Section 2.1.2. Furthermore, because we assume that the optics are lossless, the kernel satisfies

$$\int_\Omega h_s^v(y, x)dy = 1 \qquad \forall\, x \in \mathbb{R}^2, \tag{3.3}$$

which we introduced in Section 2.4 as the energy conservation principle (2.29). In the limiting case $\sigma \to 0$ in equation (3.2), when the surface is in focus, the kernel approaches Dirac's delta distribution, (also called the *identity operator*, or simply Dirac's delta),

$$h_s^v(y, x) = \delta(y - x) \tag{3.4}$$

defined by the following condition

$$I_s^r(y, v) = \int \delta(y - x)r(x)dx = r(y). \tag{3.5}$$

We now turn our attention to whether the problem of shape from defocus can be solved. In order to do so, we need to understand what "solving" means, as there are many facets to the problem. We start by assuming that the radiance can be chosen arbitrarily (as in an active illumination scenario) and ask whether shape can be recovered unambiguously, regardless of the algorithm used.

3.2 Observability of shape

Suppose we have a white Lambertian (Appendix A) scene that we can illuminate as we wish, or, even simpler, that we can directly choose its radiance. Although this is clearly an idealization, the analysis that follows is relevant to the problem of 3-D reconstruction from structured illumination.

Let us now take two or more images of this scene with different focus settings. Clearly, if these images are identical, we can say nothing about the shape of the scene, as we have noted in the example in the preamble of this chapter.

However, if we shine a narrow enough light beam that projects onto a small "dot" on the scene, its image is the point-spread function, which depends on the depth of the scene at that point. We may then actively change the focus setting to bring the point into focus, and read out the depth of that point from the focus setting. If we think of repeating the procedure and scanning the entire scene with our light beam, we can hope to recover the entire depth map. Alternatively, we can shine onto the scene a grid of points, and get a depth map at the grid. This section establishes that this procedure, in principle, allows us to distinguish two scenes that have different shape. Therefore, again in principle, it is possible to reconstruct the shape of a scene uniquely.

Consider two scenes with surfaces s and s', respectively.

Definition 3.1. *We say that a surface s is* weakly indistinguishable *from a surface s' if, for all possible radiances r, there exists at least a radiance r' such that*

$$I_s^r(y,v) = I_{s'}^{r'}(y,v) \qquad \forall\, y \in \Omega \qquad \forall\, v \in \mathcal{V}. \tag{3.6}$$

Two surfaces are weakly distinguishable *if they are not weakly indistinguishable. If a surface s is weakly distinguishable from any other surface, we say that it is* weakly observable.

In this section we establish that smooth surfaces are weakly observable.[3] In order to prove this statement we need to make some properties of an ideal imaging system explicit. In particular,

1. For each scalar $z > 0$, and for each $y \in \Omega$, there exists a focus setting v_0 and an open set $O \subset \Omega$, such that $y \in O$ and, for any surface s that is smooth in O with $s(y) = z$, we have

$$h_s^{v_0}(y,x) = \delta(y-x) \qquad \forall\, x \in O. \tag{3.7}$$

2. Given a scalar $z > 0$ and a $y \in \Omega$, such a focus setting v_0 is unique in $\mathcal{V}$.

3. For any surface s and for any open set $O' \subset \Omega$ we have

$$h_s^v(y,x) > 0 \qquad \forall\, x,\, y \in O' \tag{3.8}$$

whenever $v \neq v_0$.

The above statements formalize some of the typical properties of ideal finite aperture cameras. The first statement corresponds to guaranteeing the existence of a focus setting v_0 at each point y such that the resulting image of that point is "in focus." Notice that the focus setting v may change at each point y depending on the scene's 3-D structure. The second property states that such a focus setting is unique. Finally, when the kernel is not Dirac's delta, it is a positive function over an open set.

We wish to emphasize that the existence of a focus setting v_0 is a mathematical idealization. In practice, diffraction and other optical effects prevent a kernel from approaching a delta distribution. Nevertheless, analysis based on an idealized model can shed light on the design of algorithms operating on real data. Under the above conditions we can state the following result.

Proposition 3.2. *Any smooth surface s is weakly observable.*

Proof. See Appendix C.1. □

Proposition 3.2 shows that it is possible, in principle, to distinguish the shape of a surface from that of any other surface by looking at images under different camera settings v and radiances r. This, however, requires the active usage of the control v and of the radiance r. Although in some circumstances it may be

---

[3] We do not consider surfaces that differ by a set of measure zero.

possible to change r by projecting structured light patterns onto the scene, this is not possible in general.

The definition of weak observability leaves us the freedom to choose the radiance r, possibly as a function of s, to distinguish s from s'. However, one may wonder if a radiance r can be found that allows distinguishing any s. We address this issue in the next section.

3.3   The role of radiance

Let $\mathcal{I}(s|r)$ denote the set of surfaces that cannot be distinguished from s for a given radiance r:

$$\mathcal{I}(s|r) = \{\tilde{s} \mid I_{\tilde{s}}^r(y,v) = I_s^r(y,v) \ \ \forall \, y \in \Omega, v \in \mathcal{V}\}. \tag{3.9}$$

Note that we are using the same radiance on both surfaces $s, \tilde{s}$. This happens, for instance, when we project a static light pattern onto a surface. Clearly, not all radiances allow distinguishing two surfaces. For instance, $r = 0$ does not allow distinguishing any two surfaces. We now discuss the existence of a "sufficiently exciting" radiance that allows distinguishing any two surfaces.

Definition 3.3. *We say that a radiance distribution[4] r is* sufficiently exciting *for s if*

$$\mathcal{I}(s|r) = \{s\}. \tag{3.10}$$

The ability of a radiance r to be sufficiently exciting depends on the relationship between the kernel h_s^v and the surface s. For example, if the map $s \mapsto h_s^v$ is not one-to-one, there will be no radiance r able to retrieve the original surface s. When there is no such limitation, we conjecture that one could, in principle, construct a "universally exciting" radiance, that is, a radiance which allows distinguishing any two surfaces, but that such a radiance would have unbounded variation on any open subset of the image domain Ω. The conjecture arises from the discussion relative to square-integrable radiances that follows in Section 3.3.2, as we articulate in Remark 3.10.

Although on one hand we want to identify radiances that allow us to distinguish between any two shapes, it is also important to identify radiances that do not allow us to distinguish between any. In other words, we want to characterize radiances that are irrelevant for the purpose of 3-D reconstruction, or in other words "carry no shape information." This we do in the next section.

---

[4] We often refer to r as radiance or distribution.

3.3.1 Harmonic components

In this section, we establish that any radiance that is harmonic, or any harmonic component of a radiance, carries no shape information. Before doing so, we recall the basic properties of harmonic functions.

Definition 3.4. *A function* $r : \mathbb{R}^2 \mapsto \mathbb{R}$ *is said to be* harmonic *in an open region* $O \subset \mathbb{R}^2$ *if*

$$\triangle r(x) \doteq \sum_{i=1}^{2} \frac{\partial^2}{\partial x_i^2} r(x) = 0$$

for all $x \in O$.

Proposition 3.5 (mean-value property). *If a radiance* $r : \mathbb{R}^2 \mapsto \mathbb{R}$ *is harmonic, then, for any kernel* $h : \mathbb{R}^2 \mapsto \mathbb{R}$ *that is rotationally symmetric, we have*

$$\int h(y - x)r(x)dx = r(y) \int h(x)dx. \tag{3.11}$$

Proof. See Appendix C.2.   □

By the energy conservation principle, $\int h_s^v(y, x)\, dy = 1$; when we are imaging an equifocal plane, the kernel h_s^v is indeed rotationally symmetric and shift-invariant. The mean-value property tells us that, if r is harmonic, its image under any focus setting is the same, hence no information on shape can be obtained from defocused images.

Corollary 3.6. *If* r *is a harmonic function, then*

$$\int h_s^v(y, x)r(x)dx \simeq r(y) \tag{3.12}$$

is independent of s, *so that we can conclude that harmonic radiances do not allow distinguishing scenes with different shape.*

A simple example of harmonic radiance is $r(x) = a_0^T x + a_1$, where $a_0 \in \mathbb{R}^2$ and $a_1 \in \mathbb{R}$ are constants. It is easy to convince ourselves that when imaging an equifocal plane in real scenes, images generated by such a radiance do not change with the focus setting, as we illustrate in Figure 3.1.

Because we always measure images on a bounded domain, the set of radiances that do not carry shape information include all those that are harmonic within the same domain, such as those shown in Figure 3.1.

3.3.2 Band-limited radiances and degree of resolution

In this section we analyze the observability of shape when the radiance is fixed once and for all. In order to make the analysis simple, we are forced to make somewhat unrealistic assumptions. In particular, we allow both the radiance and

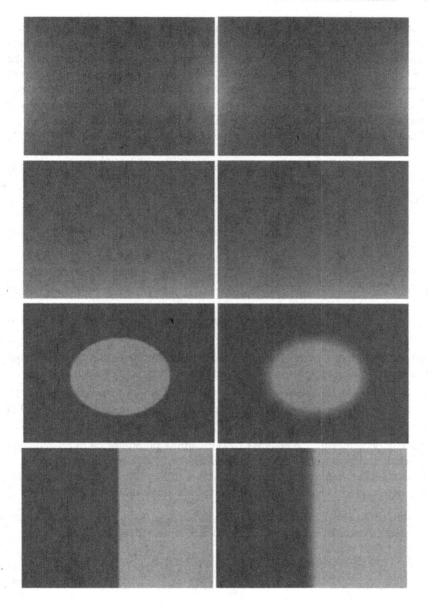

Figure 3.1. Harmonic radiances are unaffected by focus setting. We show an image of a planar scene with two patches that have harmonic radiance ($I(y_1, y_2) = y_2^2 - y_1^2$ on the top row, and $I(y_1, y_2) = y_2$ on the second row), and two patches that have a nonharmonic radiance in the third and fourth rows. When the scene is not in focus, the nonharmonic patches appear blurred, whereas the harmonic patches are unchanged.

the kernel to be square-integrable. In practice both the kernel and the radiance may violate this assumption: the kernel because it can approach a delta, and the radiance because it may be rather irregular, depending on the texture of the scene. However, assuming that $r \in L^2(\mathbb{R}^2)$ and $h_s^v(y, \cdot) \in L^2(\mathbb{R}^2)$ allows us to decompose them into orthonormal bases, which makes the analysis quite straight-forward. It is possible to relax these assumptions, but at the price of significant mathematical complications, which we wish to stay clear of in this book.

So, suppose we are given an orthonormal basis $\{\theta_k\}_{k=0...\infty}$, $\theta_k : \mathbb{R}^2 \mapsto \mathbb{R}^2$,

$$\langle\langle \theta_i, \theta_j \rangle\rangle = \int \theta_i(x)\theta_j(x)dx = \delta_{ij} = \begin{cases} 1 & if\ i = j \\ 0 & \text{otherwise} \end{cases} \qquad (3.13)$$

so that the radiance r can be written as

$$r(x) = \sum_{k=0}^{\infty} \alpha_k \theta_k(x), \qquad (3.14)$$

where $\alpha_k = \int \theta_k(x)r(x)dx$. Similarly, if the kernel h is square-integrable, we can express it as

$$h_s^v(y, x) = \sum_{k=0}^{\infty} \gamma_k(y, v, s)\theta_k(x), \qquad (3.15)$$

where the argument s reminds us of the dependency of the kernel from the shape of the scene. With this notation at hand, an image I can be written as follows

$$I_s^r(y) = \sum_{k=0}^{\infty} \gamma_k(y, v, s)\alpha_k. \qquad (3.16)$$

The above decomposition equation (3.14) identifies an infinite set of components that, when combined, give rise to the radiance r. Typically we can sort α_k in decreasing order, and truncate beyond a certain index. When the sum is actually finite, we can give a name to the maximum index in the sum.

Definition 3.7. *For a given orthonormal basis $\{\theta_k\}_{k=0...\infty}$, we say that a radiance r has a* degree of resolution ρ *if there exists a set of coefficients $\{\alpha_k\}_{k=0...\rho}$, with $\alpha_k \neq 0$, $k = 0, \ldots, \rho$, such that*

$$r(x) = \sum_{k=0}^{\rho} \alpha_k \theta_k(x). \qquad (3.17)$$

Furthermore, for a given orthonormal basis $\{\theta_k\}_{k=0...\infty}$, we say that the distribution r is band-limited *if the corresponding degree of resolution $\rho < \infty$.*

Similarly, we can define a degree of resolution for the kernel h.

Definition 3.8. *Let $h_s^v \in L^2(\mathbb{R}^2 \times \mathbb{R}^2)$. Given an orthonormal basis $\{\theta_k\}_{k=0...\infty}$, if there exists a positive real integer $\rho < \infty$ and coefficients $\{\gamma_k(y, v, s)\}_{k=0...\rho}$,* *with $\gamma_k(y,v,s)_\rho \neq 0$ for some $y \in \Omega, v \in \mathcal{V}$, such that*

$$h_s^v(y,x) = \sum_{k=0}^{\rho} \gamma_k(y,v,s)\theta_k(x) \qquad (3.18)$$

then we say that the surface s has a degree of resolution ρ.

The above definition of degree of resolution for the radiance r and the surface s depends on the choice of orthonormal basis $\{\theta_k\}$. One could make the definition independent of the choice of basis by defining ρ as the minimum integer with respect to the choice of basis in a class. Alternatively, one can just choose the same basis for both definitions, in particular when comparing different surfaces and different radiances as we show below. There is a natural link between the degree of resolution of a radiance and the degree to which two surfaces can be distinguished.

Proposition 3.9. *Let r be a band-limited distribution with degree of resolution ρ, and let $\{\theta_k\}_{k=0\ldots\infty}$ be the corresponding orthonormal basis. Then two surfaces s_1 and s_2 can only be distinguished up to the resolution determined by ρ; that is, if we write*

$$h_{s_i}^v(y,x) = \sum_{k=0}^{\infty} \gamma_k(y,v,s_i)\theta_k(x) \qquad i = 1,2 \qquad (3.19)$$

then

$$\mathcal{I}(s_1|r) \supset \left\{ s_2 \Big| \gamma_k(y,v,s_1) = \gamma_k(y,v,s_2), \forall y \in \Omega, v \in \mathcal{V}, k = 0\ldots\rho \right\}. \tag{3.20}$$

Proof. From the definition of the images $I_{s_1}^r(y)$ and $I_{s_2}^r(y)$ in equation (3.16), we have:

$$I_{s_1}^r(y) = \sum_{k=0}^{\rho} \gamma_k(y,v,s_1)\alpha_k = \sum_{k=0}^{\rho} \gamma_k(y,v,s_2)\alpha_k = I_{s_2}^r(y) \qquad (3.21)$$

which is satisfied by all the shapes in (3.20). □

Remark 3.10. *The practical value of the last proposition is to state that, the more "irregular" the radiance, the more resolving power it has. When we can project structured light on the scene, the proposition establishes that "high-frequency" patterns should be used in conjunction with shape from defocus. An "infinitely irregular" pattern should therefore be universally exciting, as conjectured in the previous section. This is why in the experiments in the next chapters we always use highly textured scenes.*

The degree of resolution with which two surfaces can be distinguished also depends upon the optics of the imaging system. In our analysis this is taken into account by the degree of complexity of the radiance, that can itself be limited, or it can be limited by the optics of the imaging system.

Now, usually we do not get to choose what the radiance of the scene is, nor its degree of resolution. In the next section we relax the definition of observability to the case when we do not know the radiance.

3.4   Joint observability of shape and radiance

Definition 3.11. *We say that the pair* (s_2, r_2) *is indistinguishable from the pair* (s_1, r_1) *if*

$$I_{s_1}^{r_1}(y, v) = I_{s_2}^{r_2}(y, v) \qquad \forall\, y \in \Omega \qquad \forall\, v \in \mathcal{V}. \qquad (3.22)$$

If the set of pairs that are indistinguishable from (s_1, r_1) is a singleton (i.e., it contains only (s_1, r_1)), we say (s_1, r_1) is *strongly distinguishable*. If we assume that both the radiance r and the kernel h are square integrable (see Section 3.3.2), it is immediate to conclude that for (s_2, r_2) to be indistinguishable from (s_1, r_1) we must have

$$\sum_{k=0}^{\min\{\rho_1, \rho_2\}} \gamma_k(y, v, s_1) \alpha_{1,k} = \sum_{k=0}^{\min\{\rho_1, \rho_2\}} \gamma_k(y, v, s_2) \alpha_{2,k} \qquad (3.23)$$

for all y and v, and where $r_1(x) = \sum_{k=0}^{\rho_1} \alpha_{1,k} \theta_k(x)$ and $r_2(x) = \sum_{k=0}^{\rho_2} \alpha_{2,k} \theta_k(x)$. It follows that we cannot draw further conclusions unless we make more stringent assumptions on the kernel $h_{s_i}^v(y, x)$. Indeed, if $h_{s_i}^v(y, x)$ is such that the vector $[\gamma_0(y, v, s_i), \ldots, \gamma_\rho(y, v, s_i)]$, with $\rho = \max\{\rho_1, \rho_2\}$, does not span the entire space $\mathbb{R}^\rho$, such as in band-limited point-spread functions, then any pair (s, r) is indistinguishable. For instance, in the simple case of $s_1 = s_2 = s$, we have

$$\sum_{k=0}^{\rho} \gamma_k(y, v, s)(\alpha_{1,k} - \alpha_{2,k}) = 0 \qquad (3.24)$$

which is satisfied whenever the difference of the two radiances lies in the null-space of $[\gamma_0(y, v, s), \ldots, \gamma_\rho(y, v, s)]$. So, there is an entire equivalence class of pairs (s, r) that generate the same image I. Therefore, we need a method to select one particular object (s, r) from the equivalence class. One way to do so is to make use of additional structure of the problem. For instance, one could impose that the object be smooth or its depth and radiance positive. This can be done through a methodology called *regularization*, which we introduce in the next section and use throughout the rest of this book.

3.5   Regularization

As mentioned in the previous section, there are infinitely many pairs (objects) (s, r) that generate the same images. This is a typical trait of so-called ill-posed problems. In order to arrive at a unique solution, assuming that the observability conditions discussed in previous sections are satisfied, we need to impose further constraints. For instance, of all possible solutions, we may wish to find the smoothest, or the one with minimum energy, or one that satisfies additional constraints. Such constraints can be imposed in a minimization scheme, as we do in the next chapters.

Regularization is a broad topic that is well beyond the scope of this book. In Appendix F we only summarize the elements that are crucial to the design of shape from defocus algorithms, and refer the reader to [Kirsch, 1996] for more details.

3.6 On the choice of objective function in shape from defocus

If we were provided with the radiance r and the shape s of a scene, we could use equation (2.9) to generate a synthetic image I. In general, this image will not be identical to the image J measured by a physical sensor. This is due to the fact that the imaging model we use is only a coarse approximation of the real image formation process. The difference between the two images can be characterized by introducing a model of their discrepancy, which we call "noise," in equation (2.9) (see Section 2.1.3). The measured image J is a realization of a complex process, part of which is modeled probabilistically using a probability density p_n. The most common choices are the Gaussian and the Poisson distributions. These can be justified in an axiomatic way [Csiszár, 1991] by defining a set of desirable properties that a discrepancy measure should satisfy. Csiszár concludes that, when the quantities involved are constrained to be positive, the only choice of criterion consistent with his axioms is the information-divergence, which we will define shortly. When there are no positivity constraints, Csiszár argues that the only consistent choice of discrepancy criterion is the least-squares distance.

We always measure images on a discrete domain, so in both cases we measure the discrepancy on $\Omega \subset Z^2$, which is reflected in the algorithms described in Chapters 4 and 5. However, in certain cases it is easier to neglect the discretization of the domain and assume that the images are defined in the continuum $\Omega \subset \mathbb{R}^2$. This is the approach we follow in Chapters 6 through 8. In this chapter, therefore, we use both characterizations of the domain.

In either case we assume independence of the noise n at each pixel $y \in \Omega$, and that the mean of n is the "ideal" image I; that is,

$$I = \int J p_n(J) dJ. \tag{3.25}$$

In the case of Gaussian noise we write the probability p_n as:

$$p_n(J) = \prod_{y \in \Omega} \frac{1}{\sqrt{2\pi\sigma^2}} e^{-(I(y)-J(y))^2/(2\sigma^2)} \tag{3.26}$$

where σ is the standard deviation of the noise n. To infer the unknown shape s and radiance r of the scene, we may maximize the log-likelihood[5]

$$\log(p_n(J)) = \sum_{y \in \Omega} -\log \sqrt{2\pi\sigma^2} - \frac{(I(y) - J(y))^2}{2\sigma^2} \qquad (3.27)$$

so that

$$\hat{s}, \hat{r} = \arg\max_{s,r} \log(p_n(J)) = \arg\min_{s,r} \sum_{y \in \Omega} (I_s^r(y) - J(y))^2 \qquad (3.28)$$

or, in the case of a continuous domain,

$$\hat{s}, \hat{r} = \arg\min_{s,r} \int_{\Omega} \|I_s^r(y) - J(y)\|^2 dy. \qquad (3.29)$$

In the former case $\sum_{y \in \Omega}(I_s^r(y) - J(y))^2$ is the ℓ^2 norm of the difference between I and J, or ℓ^2 distance, whereas in the second case the integral is the L^2 distance. In both cases we talk about a least-squares criterion, whose minimization yields the so-called *least-squares* solution.

In the case of Poisson noise we write the probability p_n as

$$p_n(J) = \prod_{y \in \Omega} \frac{I(y)^{J(y)} e^{-I(y)}}{J(y)!}. \qquad (3.30)$$

To infer the unknown shape s and radiance r of the scene, we may maximize the log-likelihood

$$\log(p_n(J)) = \sum_{y \in \Omega} J(y) \log I(y) - I(y) - \log J(y)! \qquad (3.31)$$

so that

$$\begin{aligned} \hat{s}, \hat{r} &= \arg\max_{s,r} \log(p_n(J)) \\ &= \arg\min_{s,r} \sum_{y \in \Omega} \left[J(y) \log \frac{J(y)}{I_s^r(y)} - J(y) + I_s^r(y) \right] \\ &\simeq \arg\min_{s,r} \int_{\Omega} J(y) \log \frac{J(y)}{I_s^r(y)} - J(y) + I_s^r(y) dy. \end{aligned} \qquad (3.32)$$

Notice that, in this case, the cost functional minimized in equation (3.32) is the information-divergence, or I-divergence, that generalizes the Kullback–Leibler divergence, or pseudo-distance, between two probability densities.

In summary, the inference of radiance and shape can be posed as the following cost functional minimization

$$\hat{s}, \hat{r} = \arg\min_{s,r} \Phi(J|I_s^r), \qquad (3.33)$$

---

[5]In this chapter and all subsequent chapters, log indicates the natural logarithm.

where the objective function Φ is either the ℓ^2 norm

$$\Phi(J, I_s^r) = \sum_{y \in \Omega} (I_s^r(y) - J(y))^2 \qquad (3.34)$$

or the information-divergence

$$\Phi(J|I_s^r) = \sum_{y \in \Omega} \left[J(y) \log \frac{J(y)}{I_s^r(y)} - J(y) + I_s^r(y) \right] \qquad (3.35)$$

or their counterparts integrated on a continuous domain. Naturally, one may choose a discrepancy criterion that is not tied to a probabilistic model, for instance, any L^p norm, and in particular total variation (L^1) has received a considerable amount of attention in image processing due to its amenability to be minimized with efficient numerical schemes. In the following two chapters, we present solutions to the problem of shape from defocus for the case of leastsquares (Chapter 4) and information-divergence (Chapter 5).

3.7 Summary

When we can choose the radiance of a scene, for instance, by using structured light, we can always, in principle, distinguish any two scenes that have different shape, up to a set of measure zero. This means that, at least in theory, given a scene we can always find a structured light pattern that allows us to distinguish it from any other scene. In practice this light pattern may be infinitely complex, so it may not be physically plausible, but the more complex the light pattern, the more resolving power we have in shape from defocus with structured light. Harmonic components of the radiance are useless as far as allowing us to distinguish different shapes. When we cannot choose the radiance of the scene, there is a strong link between the degree of resolution (the "complexity") of the radiance and the extent to which we can distinguish different shapes.

In practice, images are only coarsely approximated by the idealized model we developed in the previous section, so reconstruction cannot be done by "inverting" the model in a straightforward fashion. First, the unknown are infinite-dimensional, so regularization is necessary. Second, one can pose the 3-D reconstruction problem as an optimization task where one wishes to minimize the discrepancy between the actual image and the idealized model. There are many ways to measure such a discrepancy, and the most common ones are the L^2 norm and the I-divergence, corresponding to a Gaussian and Poisson model of uncertainty, respectively.

4

Least-squares shape from defocus

The problem of inferring the 3-D shape and radiance of a scene from blurred images can be posed as the minimization of a cost functional, as we have seen in Section 3.6. There we have shown that the cost functional derived from the assumption that the noise is additive and Gaussian yields a least-squares formulation of the problem of shape from defocus. This assumption is not necessarily the most realistic, but it yields a particularly simple solution that separates the estimate of shape (shape from defocus) from that of radiance (image restoration). In this chapter we explore this approach; although based on questionable assumptions, it results in particularly simple, intuitive, and instructive algorithms. The reader should revisit Section 2.1.4 where we introduce the operator notation that we use extensively in this chapter.

4.1 Least-squares minimization

In this chapter we consider the discrete image domain of $N \times M$ pixels, $\Omega \equiv \mathbb{R}^{N \times M}$. This could denote the entire image or any subset of it. We collect K images with focus settings $v_1 \ldots v_K$, and rearrange them into a vector $I \in \mathbb{R}^{N \times M \times K} \sim \mathbb{R}^W$, where $W = NMK$ following the procedure detailed in Section 2.1.4. There we have seen that we can think of the measured image J as a sample from a Gaussian density having the ideal image I as its mean. Because the density is Gaussian, the noise-free image I and and the observed image J are related by a simple additive model

$$J = I + n \tag{4.1}$$

where n is a zero-mean Gaussian noise. In other words, we have isolated the mean and called it the ideal, noise-free, image. We are now interested in finding the 3-D structure of the scene s and the radiance r that make the ideal image I, which we write as I_s^r to emphasize its dependency on the unknowns, as close as possible to the observed one. "Closeness" in this context simply means that the noise n is as small as possible when measured in the ℓ^2 sense. To summarize, we are looking for $\hat{s}, \hat{r} = \arg\min_{s,r} |n|^2$ subject to the constraint (4.1). Because of the additive nature of the noise, we can just substitute it from (4.1) to obtain the following least-squares formulation of the shape from defocus and image restoration problem

$$\hat{s}, \hat{r} = \arg\min_{s,r} |J - I_s^r|^2. \tag{4.2}$$

In Section 3.6 we have shown that this corresponds to the maximum likelihood solution relative to the model (4.1).

Now, recall the imaging model (2.14) in the notation introduced in Section 2.1.4:

$$I_s^r = H_s r. \tag{4.3}$$

The linear operator H_s is a function $H_s : L^2 \mapsto \mathbb{R}^W$, and the radiance r is a function $r : \mathbb{R}^2 \mapsto \mathbb{R}$ in L^2. The operator H_s represents the linear mapping $r \mapsto I_s^r = H_s r$ that, written explicitly, is

$$I_s^r(y) = \int h_s(y, x) r(x) dx \tag{4.4}$$

where h_s is the point-spread function, defined in Chapter 2. With this notation, the original problem (4.2) can be stated concisely as

$$\hat{s}, \hat{r} \doteq \arg\min_{s \in S, r \in L^2(\mathbb{R}^2)} |J - H_s r|^2 \tag{4.5}$$

for S a suitable finite-dimensional set of smooth functions $s : \mathbb{R}^2 \to [0, \infty)$.

This notation is not only convenient, because it is reminiscent of a familiar finite-dimensional linear system, to be solved in a least-squares sense as customary in linear algebra, but also enlightening. In fact, it allows us to use the geometry of operators between Hilbert spaces to arrive at a simple solution of the problem (4.2) that minimizes a quadratic cost function. The basic idea is to use orthogonal projections to "eliminate" the radiance, thus separating the problem of shape from defocus from that of image restoration. Because the estimate of depth can be implemented independently at each pixel, this procedure effectively reduces the problem to a one-dimensional search.

Before proceeding further we need to remind the reader of what the orthogonal projector is. Given a matrix A, its (left) orthogonal projector is another matrix $A^\perp$ such that $A^\perp A = 0$. For a semi-infinite operator (i.e., one that maps an infinite-dimensional space (such as L^2) onto a finite-dimensional one (such as $\mathbb{R}^k$)), for instance, a vector-valued integral H such as the one we deal with in this section, the orthogonal projector is a function-valued matrix $H^\perp$ such that $H^\perp H = 0$.

We can also define the Moore–Penrose pseudo-inverse $H^\dagger$ as the operator such that $H^\dagger H = 1$, the identity operator. The relationship between the pseudo-inverse and the orthogonal projector is simply $H^\perp = 1 - HH^\dagger$. Note that if the operator is invertible (e.g., a square matrix with nonzero determinant), then $H^\dagger = H^{-1}$ and the orthogonal projector is trivial. However, when the operator is not invertible, as in our case, the orthogonal projector turns out to be a useful tool.

Consider for a moment the possibility that we can find a scene (i.e., a shape and a radiance) so that $n = 0$ in (4.1), or $J = I_s^r = H_s r$. If we apply the operator $H_s^\perp$ to both sides of this equation, we obtain $H_s^\perp J = 0$, where now the shape s is the only unknown, because the radiance $r = H_s^\dagger I_s^r$ has been eliminated by the orthogonal projector. This trick is possible because the radiance r appears linearly in the image formation model. However, who guarantees that the shape s we obtain by solving the reduced problem is the same we would have obtained by solving the original problem? For the sake of analogy, consider solving a linear system of equations, say $Ax = 0$; the set of solutions is $x \in Null(A)$. However, if we multiply both sides by a singular (noninvertible) matrix B, $BAx = 0$, we are introducing a whole new set of solutions $x \mid Ax \in Null(B)$ that has nothing to do with the original problem. This is analogous to our situation, because $H_s^\perp$ is exactly a square but noninvertible matrix. Therefore, we have to pay attention that by applying this trick we are not introducing spurious solutions. This is indeed the case, as we show in the next proposition, so the reader can rest assured that no phantom solutions will emerge as a consequence of mathematical trickery.

Proposition 4.1. *Let $\hat{s}$, $\hat{r}$ be local extrema of the functional*

$$E(s,r) \doteq |J - H_s r|^2 \tag{4.6}$$

and, assuming that $H_s^\dagger$ exists, let $\tilde{s}$ be a local extremum of the functional

$$\psi(s) \doteq |H_s^\perp J|^2. \tag{4.7}$$

Furthermore, let $\tilde{r}$ be obtained from s by

$$\tilde{r} \doteq H_s^\dagger J. \tag{4.8}$$

Then $\hat{s}$ is also a local extremum of $\psi(s)$, and $\tilde{s}, \tilde{r}$ are also local extrema of $E(s,r)$.

Proof. See Appendix C.3. □

Remark 4.2. *The significance of the proposition above is that equation (4.6) and equation (4.7) admit the same 3-D structure as a minimizer but, whereas equation (4.6) is a cost functional in two unknowns, equation (4.7) is a cost functional in a single unknown, s. Furthermore, if we constrain surfaces s to belong to a finite-dimensional set, and equation (4.6) leads to an infinite-dimensional problem, equation (4.7) leads to a finite-dimensional one.*

This seems too good to be true, and indeed there are caveats and details that we have swept under the rug. We have transformed an infinite-dimensional problem into a finite-dimensional one, finding the depth at each pixel. The danger here is in the fine print, that is, the assumption, in the previous proposition, that the pseudo-inverse exists. This, as we show, is equivalent to a regularity assumption on r. If $H_{\hat{s}}$ is injective[1] for a given $\hat{s}$, the orthogonal operator $H_{\hat{s}}^{\perp}$ is the null map (see equation (2.18)). In this case, equation (4.7) is trivially satisfied for any measured image I, and therefore $\hat{s}$ is always a minimizer. Hence, a necessary condition to avoid this scenario is to impose that $H_{\hat{s}}$ maps functions in $L^2(\mathbb{R}^2)$ to a proper subspace of $\mathbb{R}^W$ of dimension less than W. In the next section, we do so by truncating the singular values of either the operator H_s or the operator $H_s^{\perp}$, which corresponds to assuming that such operators are band-limited.

As we have seen in the proposition, rather than solving the original problem in equation (4.6), we can solve the simpler problem in equation (4.7). Then, the minimization of equation (4.6) boils down to computing the orthogonal operators $H_s^{\perp}$. As we show in the next section, $H_s^{\perp}$ can be computed in different ways.

4.2 A solution based on orthogonal projectors

When the PSF of the camera is known, one can directly compute the orthogonal operators $H_s^{\perp}$ in closed form, at least for simple classes of PSFs. More in general, one can express H_s via the functional singular value decomposition [Bertero and Boccacci, 1998], as we discuss in Section 4.2.1. When the PSF is not known, one can compute $H_s^{\perp}$ directly from blurred images, as we explain in Section 4.2.2. The advantage of this second solution is its simplicity. To compute $H_s^{\perp}$ one only needs to collect a training set of controlled blurred images, and then express the training set via the singular value decomposition.

4.2.1 Regularization via truncation of singular values

Assuming that we have the exact model of the PSF, we can express the operator H_s using its functional singular value decomposition. Let $\{\lambda_k \geq 0\}_{k=1,...,W}$ be a sequence of nonnegative scalars sorted in decreasing order, $\{v_k\}_{k=1,...,W}$ an orthonormal set of vectors in $\mathbb{R}^W$ and $\{u_k\}_{k=1,...,W}$ an orthonormal set of functions[2] in $L^2(\mathbb{R}^2)$. Then, H_s can be written as

$$H_s = \sum_{k=1}^{W} \lambda_k u_k v_k, \tag{4.9}$$

where H_s maps $L^2(\mathbb{R}^2)$ onto $\mathbb{R}^W$ as follows

$$r \mapsto H_s r \doteq \sum_{k=1}^{W} \lambda_k \langle r, u_k \rangle v_k. \tag{4.10}$$

---

[1]Recall that a function is injective when its range is the whole co-domain. In our case, $H_{\hat{s}}$ is injective when for each image $I \in \mathbb{R}^W$ there exists a radiance r such that $H_{\hat{s}} r = I$.

[2]Notice that there can be at most W orthonormal vectors $v_k \in \mathbb{R}^W$, and therefore, all the singular values λ_k and functions u_k relative to the same vector v_k can be lumped together.

Using the same sequences of u_k and v_k, we can obtain an expression for the adjoint H_s^*. The adjoint is defined by the equation

$$H_s^* I \doteq \sum_{k=1}^{W} \lambda_k \langle\!\langle I, v_k \rangle\!\rangle u_k. \tag{4.11}$$

It is easy to verify by substitution that the pseudo-inverse is given by

$$H_s^\dagger = \sum_{k=1}^{\rho} \lambda_k^{-1} u_k v_k^T, \tag{4.12}$$

where $\rho \leq W$ is a suitable integer (the rank of the operator), and $\lambda_k > 0 \,\forall\, k \leq \rho$. Then, the orthogonal projection operator is

$$H_s^\perp = \mathbf{1} - \sum_{k=1}^{\rho} v_k v_k^T, \tag{4.13}$$

where $\mathbf{1}$ is the $W \times W$ identity matrix. In order for the orthogonal projection operator to be nontrivial, we need to assume that $\rho < W$. This is equivalent to assuming that H_s maps to a proper subspace of $\mathbb{R}^W$, which imposes a lower bound on the dimensionality of the data to be acquired, that is, the minimum number of blurred images and their size.

The sequences $\{\lambda_k\}$, $\{u_k\}$, and $\{v_k\}$ are found by solving the normal equations:

$$\begin{cases} H_s^* H_s u_k = \lambda_k^2 u_k \\ H_s H_s^* v_k = \lambda_k^2 v_k \end{cases} \quad k = 1 \ldots \rho \tag{4.14}$$

or, making the notation explicit

$$\begin{cases} \sum_y h^s(\tilde{x}, y) \left(\int h^s(x, y) u_k(x) dx \right) = \lambda_k^2 u_k(\tilde{x}) \\ \int h^s(x, \tilde{y}) \left(\sum_y h^s(x, y) v_k(y) \right) dx = \lambda_k^2 v_k(\tilde{y}) \end{cases} \quad k = 1 \ldots \rho. \tag{4.15}$$

The second of the normal equations (4.15) can be written as

$$\mathcal{M} v_k = \lambda_k^2 v_k \quad k = 1 \ldots \rho \tag{4.16}$$

where $\mathcal{M}$ is a W-dimensional symmetric matrix with elements $\langle h^s(\cdot, \tilde{y}), h^s(\cdot, y) \rangle$. Because this is a (finite-dimensional) symmetric eigenvalue problem, there exists a unique decomposition of $\mathcal{M}$ of the form

$$\mathcal{M} = V \Lambda^2 V^T \tag{4.17}$$

with $V^T V = \mathbf{1}$, $\Lambda^2 = \text{diag}\{\lambda_1^2 \ldots \lambda_\rho^2\}$, and $V = [v_1, \ldots, v_\rho]$. We are now left with the first equation in (4.15) in order to retrieve $u_k(x)$. However, instead of solving that directly, we use the adjoint operator H_s^* to map the basis of $\mathbb{R}^W$ onto a basis of a ρ-dimensional subspace of $L^2(\mathbb{R}^2)$ via $H_s^* v_k = \lambda_k u_k$. Making the

Table 4.1. Algorithm to compute the orthogonal operator $H_s^{\perp}$ with known point-spread function.

Algorithm (computation of $H_s^{\perp}$ with known PSF)
---

1. Given: Calibration parameters (from knowledge of the camera; see Appendix D for a simple calibration procedure) $\{v_i\}_{i=1,\ldots,K}, F, D, \gamma$, and the rank ρ of the orthogonal operator $H_s^{\perp}$,

2. Choose a synthetic surface s,

3. Compute the W-dimensional symmetric matrix $\mathcal{M}$ by evaluating $\langle h^s(\cdot, \tilde{y}), h^s(\cdot, y) \rangle, \forall \tilde{y}, y \in \Omega$,

4. Decompose $\mathcal{M}$ via the singular value decomposition as shown in equation (4.17); this yields the matrix $V = [v_1, \ldots, v_\rho]$,

5. Build $H_s^{\perp}$ as in equation (4.13).

notation explicit we have

$$u_k(x) = \lambda_k^{-1} \sum_y h^s(x, y) v_k(y) \qquad k = 1 \ldots \rho. \qquad (4.18)$$

The method that we have just described is summarized in Table 4.1.

Remark 4.3 (Regularization). *In the computation of $H_s^{\perp}$, the sum is effectively truncated at $k = \rho < W$, where the dimension W depends upon the amount of data acquired. As a consequence of the properties of the singular value decomposition (SVD), the solution obtained enjoys a number of regularity properties, further discussed in Appendix F, Section F.3.1. Note that the solution is not the one that we would have obtained by first writing r using a truncated orthonormal expansion in $L^2(\mathbb{R}^2)$, then expanding the kernel h^s in (2.14) in series, and then applying the finite-dimensional version of the orthogonal projection theorem.*

4.2.2 Learning the orthogonal projectors from images

When the model of the PSF is not known, we cannot use the method described in the previous section to compute the orthogonal operators. Here we show that complete knowledge of the point-spread function is indeed not necessary. To compute the orthogonal operators one only needs the finite-dimensional range of the PSF, which can also be obtained directly from a collection of blurred images.

Recall that a multifocal vector image I can be written as $H_{\bar{s}}r$ for some surface $\bar{s}$ and a radiance r (see equation (2.14) for how to stack multiple blurred images into a vector). By definition, if we multiply the orthogonal operator $H_{\bar{s}}^{\perp}$ by I, we obtain

$$H_{\bar{s}}^{\perp} I = H_{\bar{s}}^{\perp} H_{\bar{s}} r = 0. \qquad (4.19)$$

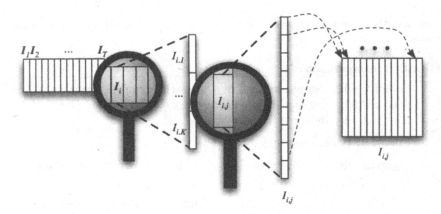

Figure 4.1. Graphic visualization of the matrix $\mathcal{I} = [I_1\ I_2\ \ldots\ I_t]$. The *ith* column of $\mathcal{I}$ is a multifocal image generated by a single radiance r_i. A multifocal image is then a collection of K defocused images captured with different focus settings. Finally, each defocused image is rearranged columnwise into a column vector.

Notice that this equation is satisfied for any radiance r. Hence, if we collect a set of t images[3] $\{I_i\}_{i=1\ldots t}$ by letting the radiance $r = \{r_1, \ldots, r_t\}$ we obtain

$$H_{\bar{s}}^{\perp}[I_1\ \ldots\ I_t] = H_{\bar{s}}^{\perp}H_{\bar{s}}[r_1\ \ldots\ r_t] = [0\ \ldots\ 0] \doteq \mathbf{0} \qquad (4.20)$$

as long as the shape $\bar{s}$ of the scene remains the same. We can therefore find $H_{\bar{s}}^{\perp}$ by simply solving the following system of linear equations.

$$H_{\bar{s}}^{\perp}[I_1\ I_2\ \ldots\ I_t] = \mathbf{0}. \qquad (4.21)$$

Notice, however, that $H_s^{\perp}$ is not a generic matrix but, rather, has some important structure that must be exploited in solving the system of equations above. In particular, $H_s^{\perp}$ is a *symmetric* matrix (i.e., $H_s^{\perp} = (H_s^{\perp})^T$) which is also *idempotent* (i.e., $H_s^{\perp} = (H_s^{\perp})^2$). According to the first property, we can write $H_s^{\perp}$ as the product of a matrix A of dimensions $m \times n$, $m \geq n$, with its transpose; as for the second property, we have that the columns of A must be orthonormal, and thus $H_s^{\perp}$ can uniquely be written as

$$H_s^{\perp} = AA^T, \qquad (4.22)$$

where $A \in V_{n,m}$ and $V_{n,m}$ is the set of rectangular matrices with m orthonormal columns.

Let $\mathcal{I} = [I_1\ I_2\ \ldots\ I_t] \in \mathbb{R}^{W \times t}$; then the solution of equation (4.21) can be obtained via the singular value decomposition of $\mathcal{I}$:

$$\mathcal{I} = UBQ^T, \qquad (4.23)$$

---

[3]Recall that each image $I_i \in \mathbb{R}^W$, $W = M \times N \times K$, is a column vector collecting K defocused images $[I_1^T \ldots I_K^T]^T$, $I_j \in \mathbb{R}^{M \times N}\ \forall\, j = 1, \ldots, K$, captured for K different focus settings.

Table 4.2. Learning algorithm to compute the orthogonal operator $H_s^\perp$ without knowledge of the point-spread function.

Algorithm (computation of $H_s^\perp$ with unknown PSF)

1. Choose a surface s, either a physical surface or a synthetic one. Generate (in the case of synthetic data) or collect (in the case of real data) a number t of "training" images $[I_1\ I_2\ \ldots\ I_t] \in \mathbb{R}^{W \times t}$ from a scene with shape s and corresponding radiances $r_1, r_2, \ldots, r_t$. To make the terms explicit, notice that for each radiance $r_j, j = 1, \ldots, t$, we have K defocused images $I_{i,j} \in \mathbb{R}^{M \times N}, i = 1, \ldots, K$ taken with K focus settings. The images $I_{i,j}$ are rearranged as a column vector and then stacked on top of each other, so as to form a column vector $I_j \in \mathbb{R}^W$, with $W = M \times N \times K$. Finally, $[I_1\ I_2\ \ldots\ I_t]$ is a matrix in $\mathbb{R}^{W \times t}$ (see Figure 4.1),

2. Collect all rearranged column vectors into the matrix $\mathcal{I} = [I_1\ I_2\ \ldots\ I_t]$ of dimensions $W \times t$. Compute the SVD of $\mathcal{I}$ so that $\mathcal{I} = UBQ^T$ (see equation (4.23)),

3. Determine the rank ρ of $\mathcal{I}$, for example, by imposing a threshold on the singular values.

4. Decompose U as $U = [U_1\ U_2]$, where U_1 contains the first ρ columns of U and U_2 the remaining columns; then build $H_s^\perp$ as

$$H_s^\perp = U_2 U_2^T.$$

where U is a $W \times W$ matrix with orthonormal column vectors, B is a $W \times t$ matrix where the upper $W \times W$ matrix is diagonal with positive values, and Q is a $t \times t$ matrix with orthonormal column vectors; the solution of equation (4.21) is then

$$H_{\bar{s}}^\perp = U_2 U_2^T, \tag{4.24}$$

where $U = [U_1\ U_2]$ and U_2 are the orthonormal vectors corresponding to the null singular values of B. In other words, given a scene with shape $\bar{s}$, we can "learn" the corresponding orthogonal operator $H_{\bar{s}}^\perp$ by applying the singular value decomposition (SVD) to a matrix whose column vectors are defocused images. In the presence of deviations from the ideal model of equation (4.21), this yields the least-squares estimate of $H_s^\perp$, which can be thought of as a learning procedure, summarized in Table 4.2.

Remark 4.4 (Excitation). *The computation of the orthogonal operator $H_s^\perp$ depends strongly on the training sequence of defocused images that we choose (see equation (4.21)). In order to be able to learn a nontrivial orthogonal operator, we expect the training set to span a subspace of dimension less than W. However, there are two factors that determine the rank of the training set. One is the intrinsic structure of the PSF, which is what we want to characterize, the other is* *the rank of the chosen set of radiances $r_1, \ldots, r_T$. For example, clearly if we use constant radiances, $r_i = $ const., we cannot recover the correct operator. If we employ radiances that are linearly dependent and span a subspace of dimension $\rho < W$, the corresponding defocused images are also linearly dependent (due to the linear relation between radiances and defocused images) and span a subspace of dimension less than or equal to ρ (some radiances may belong to the null space of H_s). If the rank due to the intrinsic structure of the PSF is larger than ρ, by using these images we do not reconstruct the correct $H_s^\perp$. To determine the correct structure of $H_s^\perp$ we need to guarantee that only the first factor is lowering the rank. In order to do that, we need to choose a set of radiances that is large enough (i.e., $T \geq W$), spans a subspace of dimension larger than or equal to W, and does not belong to the null space of H_s. We call such radiances* sufficiently exciting *for the training sequence. In practice, a good choice of radiance is a "uniform noise image," for instance, as generated by the MATLAB® command* I = 256*rand(N,M).

4.3   Depth-map estimation algorithm

So far we have seen that the original problem in equation (4.6) can be reduced to the problem in equation (4.7), which involves the computation of orthogonal projectors $H_s^\perp$ (Section 4.1). Then, we have shown two methods to compute the projectors (Section 4.2.1 and Section 4.2.2). Now that we have derived all the necessary components, we introduce a simple algorithm to reconstruct the geometry of a scene from defocused images.

Given a collection of K defocused images $I = [I_1^T \ldots I_K^T]^T$, we want to estimate the 3-D structure of the scene by minimizing the following cost functional

$$\tilde{s} = \arg\min_s \|H_s^\perp I\|^2. \tag{4.25}$$

If one is also interested in reconstructing the radiance (i.e., deblurring the defocused images I), the solution $\tilde{s}$ of equation (4.25) can be used to compute:

$$\tilde{r} = H_{\tilde{s}}^\dagger I. \tag{4.26}$$

In principle, the above minimization can be carried out by using the tools of calculus of variations as we do in the next chapters. Using the tools described in Appendix B one can estimate $\tilde{s}$ by implementing the following gradient descent flow;

$$\frac{\partial s}{\partial k} = -\nabla_s E, \tag{4.27}$$

where k is the iteration time, $E(s) = \|H_s^\perp I\|^2$ is the cost functional, and $\nabla_s E$ is the functional derivative of E with respect to the surface s.

In this chapter, however, we are interested only in the simplest implementation. Hence, rather than implementing computationally expensive gradient flows, we

Table 4.3. Least-squares algorithm to compute a solution of shape from defocus via the minimization described in Section 4.3.

Algorithm (least-squares)

1. Compute the orthogonal operator $H_s^\perp$ for a set of equifocal planes s (e.g., with depths $z_1, \ldots, z_{50}$) either by employing the algorithm described in Table 4.1 or the algorithm described in Table 4.2.

2. Given: A multifocal image $J = [J_1^T, \ldots, J_K^T]^T$, where K is the number of focus settings,

3. Locally at each pixel x of the input images,

   - Extract the patch J_{W_x}, and evaluate $\|H_s^\perp J_{W_x}\|$ for all the equifocal surfaces s,
   - Retain the depth corresponding to the minimum of $\|H_s^\perp J_{W_x}\|$ over all the equifocal surfaces s.

solve equation (4.25) locally around patches of the defocused images, and for each patch we assume that the corresponding surface belongs to a finite-dimensional set of admissible surfaces S. In particular, in the simplest case we assume that the scene that generates a small patch can be locally approximated with a plane parallel to the image plane (the equifocal assumption). In other words, we solve

$$\tilde{s}(x) = \arg\min_{s \in S} \|H_s^\perp I_{W_x}\|^2, \tag{4.28}$$

where W_x is a region centered in x, I_{W_x} is the restriction of I to W_x, and S is a discrete set of depths, for instance, $z_1, \ldots, z_{50}$. This simplification allows us to precompute the orthogonal projectors $H_s^\perp$, one for each depth level, and minimize the cost functional by a simple one-dimensional exhaustive search. The algorithm is summarized in Table 4.3 and the corresponding MATLAB® implementation is given in Appendix E, Section E.1.

This algorithm enjoys a number of properties that make it suitable for real-time implementation. First, the only operations involved are matrix-vector multiplications, which can be easily implemented in hardware. Second, the process is carried out at each pixel independently, thus enabling efficient parallel implementations. It would be possible, for example, to have CCD arrays where each pixel is a computing unit returning the depth relative to it. Also, one could compensate for the coarseness of choosing a finite set of depth levels by interpolating the computed cost function. The search process can be accelerated by using well-known descent methods (i.e., gradient descent, Newton-Raphson, tangents, etc.) or by using a dichotomic search. Notice that, despite its simplicity, the algorithm is very general, and the choice of working locally at patches is not vital to the feasibility of the algorithm. In the next section we show some examples of reconstruction results obtained with the algorithm thus described.

4.4   Examples

In this section we present a set of experiments both on synthetic data (unaffected by noise) and on real images (affected by sensor noise). Although there are many possible choices of kernels, in practice the differences in reconstruction by using a Gaussian or a pillbox point-spread function are negligible, so we use what is most convenient from a mathematical standpoint, and that is the Gaussian family. This is not a binding choice, however, because the same algorithm can be carried out for any other family of PSFs, at the cost of an increased computational burden.

Due to the discretization implicit in the sensor reading, the point-spread function we obtain is never close to Dirac's delta. Instead, it is closer to a smooth function even when the scene is brought into sharp focus. For simplicity, we explicitly model discretization via the integral of the ideal Gaussian kernel over a square pixel surface

$$\tilde{I}(y) = \int_{y_1-\Delta y/2}^{y_1+\Delta y/2} \int_{y_2-\Delta y/2}^{y_2+\Delta y/2} \int h^s(x, \tilde{y}) r(x) dx d\tilde{y}, \qquad (4.29)$$

where $y = [y_1, y_2]^T$ and $\Delta y \times \Delta y$ is the area of a pixel.

In all the experiments we compute orthogonal operators locally on windows of 7×7 or 9×9 pixels. Because we always collect two defocused images only, while changing the focus settings, the orthogonal operators are matrices of size $2 \cdot 7^2 \times 2 \cdot 7^2 = 98 \times 98$ pixels or $2 \cdot 9^2 \times 2 \cdot 9^2 = 162 \times 162$ pixels. As we show, a qualitative comparison of the reconstruction of shape from real images does not reveal a noticeable difference among the orthogonal operators computed via the method in Section 4.4.1 or the method in Section 4.2.2. In other words, we can "learn" the orthogonal projector $H_s^\perp$ corresponding to "virtual," or synthetic, cameras, and then use it to infer the shape of scenes captured with real cameras. The results are comparable with those obtained by inferring the orthogonal projector through a careful calibration procedure as we describe in Section 4.4.1. This shows the remarkable flexibility of such a simple algorithm for shape from defocus; its implementation can be found in Appendix E.

4.4.1   Explicit kernel model

Following the procedure presented in Section 4.4.1 we compute a set of orthogonal operators $H_s^\perp$ in the case of a Gaussian kernel for patches of 7×7 pixels. We simulate a scene made of 51 equifocal planes placed equidistantly in the range between 520 mm and 850 mm in front of a camera with a 35 mm lens and F-number 4. We capture two defocused images. One is obtained by bringing the plane at depth 520 mm into focus. The other is obtained by bringing the plane at depth 850 mm into focus. Each of the 51 equifocal planes corresponds to one orthogonal operator. We would like to stress that the orthogonal operators do not need to be computed for equifocal planes, but can be computed for any other set of surfaces (Section 4.3).

Once the orthogonal projectors are computed, we apply them on both real and synthetically generated images. In the synthetic case we simulate 50 defocused images for each of the 51 equifocal planes used to generate the orthogonal operators. Each of the 50 simulations is obtained by employing a radiance of the scene that is generated randomly. At each depth level, and for each of these experiments, we estimate a depth level. Figure 4.2 shows the depth estimation performance when we use the computed orthogonal operators with ranks 40, 55, 65, 70, 75, and 95. Both mean and standard deviation (in the graphs we show three times the computed standard deviation) of the estimated depth (solid line) are plotted over the ideal characteristic curve (the diagonal dotted line). Clearly, when the chosen rank does not correspond to the true rank of the operators, the performance rapidly degenerates. In this case, the correct rank is 70. For this choice the maximum estimation error[4] is 31 mm. We also test the performance of this algorithm on the real images shown in Figure 4.3 and Figure 4.6 by working on patches of 9 × 9 pixels. In Figure 4.3 the scene is composed of objects placed between 640 mm and 750 mm in front of the lens. From the bottom to the top we have: a box, a slanted plane, and two cylinders. We capture images using an eight-bit camera containing two independently moving CCDs (kindly provided to us by S. K. Nayar of Columbia University). The lens is a 35 mm Nikon NIKKOR with F-number 4. In Figure 4.4 we show the estimated depth map as a gray-level image, where light intensities correspond to points that are close to the camera, and dark intensities to points that are far from the camera. In Figure 4.5 we show the estimation results as texture-mapped surfaces, after smoothing. In Figure 4.7 we show the estimated depth map obtained from the two input images in Figure 4.6 as a gray-level image.

4.4.2 Learning the kernel model

In this section, we evaluate the performance of the depth estimation algorithm when the orthogonal operators are computed via the procedure described in Section 4.2.2. As in the previous section, we perform experiments on both real and synthetic data. We use operators computed from synthetic data on both real and synthetic imagery, and operators computed from real data on real imagery obtained from the same camera. We divide the range between 520 mm and 850 mm in front of the camera into 51 intervals, and compute 51 orthogonal projectors each corresponding to a plane parallel to the image plane placed at one of the intervals. Each operator is computed by capturing only two defocused images of 640 × 480 pixels. We collect 200 patches of 7 × 7 pixels or 9 × 9 pixels from these images and use them to estimate the orthogonal operator. We test that the radiances collected from each patch are sufficiently exciting (see Remark 4.4) by

---

[4]We compute the estimation error as the absolute value of the difference between the estimated depth of the scene and the ground truth.

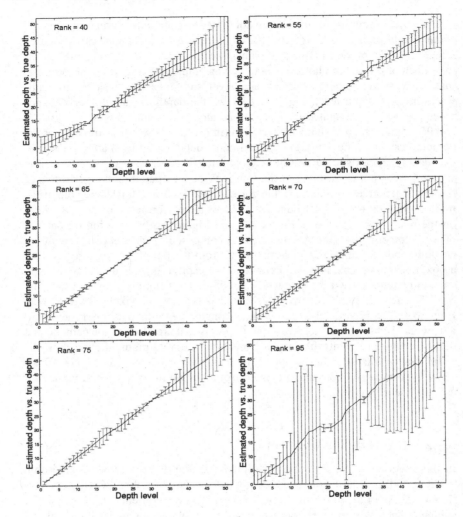

Figure 4.2. Performance test for different ranks of the orthogonal operators computed in closed form in the case of a Gaussian PSF. From left to right and from top to bottom, the ranks are 40, 55, 65, 70, 75, and 95. Both mean and standard deviation (in the graphs we show three times the computed standard deviation) of the estimated depth (solid line) are plotted over the ideal characteristic curve (dotted line) for 50 experiments.

making sure they are not linearly dependent. The whole procedure can be easily performed both on synthetic and real data.

Experiments on synthetic data.   We simulate the same scene and camera settings as in the previous section. Figure 4.8 shows the depth estimation performance when we use the orthogonal operators learned from synthetic data, with ranks 40, 55, 65, 70, 75, and 95 on synthetic images. Both mean and standard

Figure 4.3. Left: Setup of the real scene. From the bottom to the top we have: a box (parallel to the image plane), a slanted plane, and two cylinders. Right: Two images captured with different focal settings.

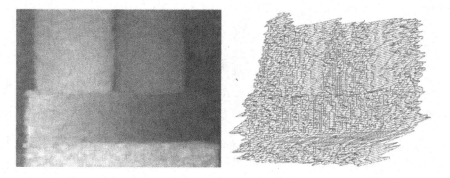

Figure 4.4. Estimated depth of the real images in Figure 4.3 when using the Gaussian PSF and the closed form solution for the orthogonal operators. Left: The estimated depth in gray-level intensities, where light corresponds to small depths, and dark to large depths. Right: Visualization of the estimated depth as a mesh.

deviation (in the graphs we show three times the computed standard deviation) of the estimated depth (solid line) are plotted over the ideal characteristic curve (the diagonal dotted line). Clearly, when the chosen rank does not correspond to the correct rank of the operators, the performance degrades rapidly. In this case, the correct choice for the rank is again 70. For this choice, the maximum estimation error is 27 mm.

Figure 4.5. Novel views of the estimated shape (after smoothing) and with texture mapping. Shape has been reconstructed with orthogonal operators computed in closed form in the case of a Gaussian PSF.

Figure 4.6. Two real images captured with different focus settings. For more details on the scene and camera settings, see [Watanabe and Nayar, 1998].

Experiments on real data.   We perform three experiments with real data. In two experiments we use operators learned from synthetic images, by assuming the PSF is Gaussian. The operators are computed as described in the previous section, applied to the pair of real images shown in Figure 4.3 and return the depth map estimate in Figure 4.9. As one can observe, the estimated depth map is very similar to the one obtained when the operators are computed in closed form, as prescribed in Section 4.2.1 (compare to Figure 4.4). We also apply these operators (learned via synthetic images) to the pair of real images shown in Figure 4.6. The estimated depth map is shown in Figure 4.10. In the third experiment we learn the orthogonal projection operators from a collection of real images and then apply these operators to novel real images obtained from the same camera.

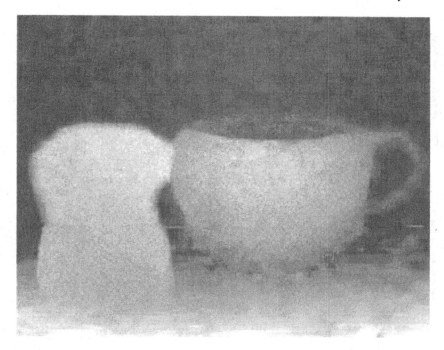

Figure 4.7. Estimated depth maps from the two input images in Figure 4.6. The depth map has not been postprocessed. We show the depth map estimated with the simple algorithm described in this chapter with known PSF.

We test these operators on the images in Figure 4.3, and obtain the depth map shown in Figure 4.11. As one can see, the estimated depth map is very similar to the one obtained in the previous experiments.

4.5 Summary

Shape from defocus can be formulated as an optimization problem. Among the possible choices of optimization criterion, the L^2 norm is attractive because of the simplicity of the resulting algorithms. The least-squares solution of shape from defocus is particularly simple because one can exploit orthogonal projections to "eliminate" the unknown radiance thus separating the problem of shape from defocus from that of image restoration.

Given some generic regularity assumptions (either the radiance or the point-spread function is band-limited), one can transform the infinite-dimensional problem of reconstructing 3-D shape and radiance into the finite-dimensional problem of computing depth at each pixel. This is done through the construction of the orthogonal operator of the linear image formation model. Such an operator can be constructed directly from simple classes of point-spread functions, or

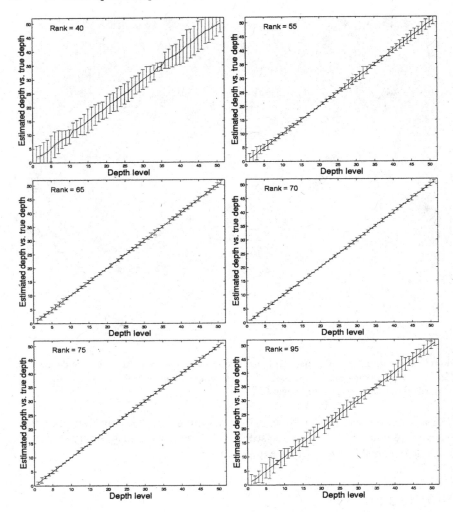

Figure 4.8. Performance test for different ranks of the orthogonal operators learned from synthetic data in the case of a Gaussian PSF. From left to right and top to bottom, the ranks are 40, 55, 65, 70, 75, and 95. Both mean and standard deviation (in the graphs we show three times the computed standard deviation) of the estimated depth (solid line) are plotted over the ideal characteristic curve (dotted line), for 50 experiments.

can be learned from data, by capturing a sufficient number of blurred images of a calibration target.

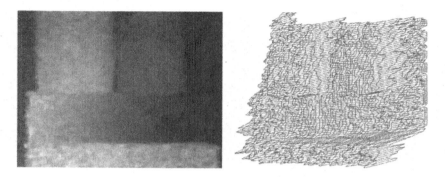

Figure 4.9. Estimated depth of the scene shown in Figure 4.3 when using the Gaussian PSF and the learning method for the orthogonal operators on synthetic images. Left: The estimated depth in gray-level intensities, where light corresponds to small depths, and dark to large depths. Right: Visualization of the estimated depth as a mesh.

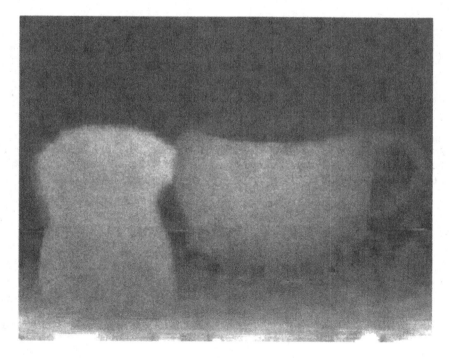

Figure 4.10. Estimated depth maps from the two input images in Figure 4.6. The depth map has not been postprocessed. We show the depth map obtained with the simple least-squares algorithm with unknown PSF.

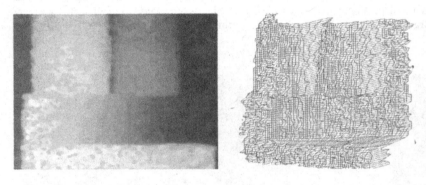

Figure 4.11. Estimated depth for the scene in Figure 4.3 when using the orthogonal operators learned from real images. Left: The estimated depth in gray-level intensities, where light corresponds to small depths, and dark to large depths. Right: Visualization of the estimated depth as a mesh.

5

Enforcing positivity: Shape from defocus and image restoration by minimizing I-divergence

The solution to the problem of recovering shape from defocus described in the previous chapter is based on the assumption that the discrepancy between the ideal image and the measured one is additive Gaussian noise. This assumption is clearly not germane, because it admits the possibility that the measured image is negative. In fact, given a large enough variance of the noise n, even if the ideal image I is positive, one cannot guarantee that the measured image $J = I + n$ is positive as well. The Gaussian assumption is desirable because it yields a least-squares solution that is particularly simple, by allowing the separation of the problem of shape from defocus from that of image restoration.

In this chapter we explore a different model, one that can guarantee that all the quantities at play, ideal image and measured one, are positive. We start by modeling noise as a Poisson, rather than Gaussian, random process. Alternatively, one can take a more formal approach that we have hinted at in Section 3.6; in either case one arrives at information-divergence, or I-divergence, as a choice of discrepancy measure, rather than least squares. We therefore pose the problem of simultaneous shape from defocus and image restoration as that of minimizing the I-divergence between the measured image and the ideal one. Now, I-divergence is not a distance, and we do not have the structure of Hilbert spaces available, so we are not able to exploit orthogonal projections and reduce the problem to linear algebra as we did in the previous chapter to find the least-squares solution. Enforcing positive depth requires us to introduce the basics of constrained optimization, and forces us to forgo a closed-form solution. Instead, we derive an iterative scheme that, under suitable conditions, converges to a desirable estimate of shape and radiance. We try to keep the discussion self-contained with the aid of Appendix E, Section E.2, where an implementation of the algorithm derived here can be used to help the reader to better understand it.

The reader may want to review Section 3.6 briefly before moving on to the next section.

5.1   Information-divergence

As in previous chapters, J denotes the measured multifocal image composed of $K \geq 2$ defocused images, whereas I is the ideal one. We assume that $\Omega \subset \mathbb{R}^2$, and that the kernel h_s^v belongs to a family of positive functions, such as Gaussians with varying standard deviation σ that relates to the focus setting as we have described in equation (2.28). We let $r : \Omega \to [0, \infty)$ be the radiance of the scene and $s : \Omega \to (0, \infty)$ be a function that represents the depth of the scene. Furthermore, we assume that s and r, our unknowns, are such that the estimated image I_s^r takes on bounded values for any $y \in \Omega$. In this case, we say that s and r are *admissible*. Note that the unknowns are defined so that they generate images that are strictly positive. [1] The problem of simultaneously estimating the shape s and the radiance r of a scene can be posed as that of finding $\hat{s}$ and $\hat{r}$ that minimize the I-divergence between the measured image and the one generated by the model, as we have shown in Section 3.6. The rationale is that, among all possible models of the scene, we seek one that assigns the measured images the highest possible likelihood under the chosen noise model. When the noise is distributed according to a Poisson distribution, maximizing the likelihood of the measured images is equivalent to minimizing the I-divergence between them and the images generated by the model, as we have shown in equation (3.32). This rationale may seem rather conservative. Even if the model of the 3-D geometry and radiance of the scene we have estimated generates images that closely match the measured ones, there is no guarantee that the 3-D geometry and radiance themselves match those of the actual scene (see Remark 5.2), nor that the estimates are unique. Chapter 3 was devoted to this problem, so in this chapter we move on to describe an actual algorithm to reconstruct 3-D shape and radiance under this model. The information-divergence Φ between two multifocal images J and I_s^r is defined as

$$\Phi\left(J | I_s^r\right) \doteq \sum_{i=1}^{K} \int_{\Omega} J(y, v_i) \log \frac{J(y, v_i)}{I_s^r(y, v_i)} - J(y, v_i) + I_s^r(y, v_i) dy, \quad (5.1)$$

---

[1] We consider a function or a value a to be *positive* if $a \geq 0$, and *strictly positive* if $a > 0$. We need ideal and measured images to be strictly positive in order for the I-divergence to be finite, as we show shortly. We enforce such an assumption by considering two cases: either the radiance is strictly positive and the kernels are positive (which includes Dirac deltas), or the radiance is positive and the kernels are strictly positive (which excludes Dirac deltas). In particular, in the case when either the radiance or the kernel is constrained to be only positive, we require such a function to be not zero everywhere (otherwise the solution is trivial). We carry out the analysis in the case of a positive radiance and strictly positive kernel, because the same results apply to the other case as a corollary.

where log indicates the natural logarithm. Notice that the function Φ above is always positive and is 0 if and only if $J(y, v_i)$ is equal to $I_s^r(y, v_i)$ at every pixel $y \in \Omega$ and for all focus settings v_i, $i = 1, \ldots, K$. However, it is not a distance because it does not satisfy the triangular inequality. Notice that it is not even symmetric, in that the I-divergence between I_s^r and J is not equal[2] to that between J and I_s^r.

Once we agree to minimize the discrepancy (5.1), we are confronted with the fact that the unknowns (the radiance r and the depth map s) are infinite-dimensional, whereas our input data, the images J, are finite-dimensional. Therefore, there are infinitely many scenes that yield images which have the same discrepancy from the measured ones. Among all of these, one could choose one that is "simplest" in some sense. The notion of simplicity, which formalizes Ockham's razor principle, can be formalized in many ways [Akaike, 1976; Rissanen, 1978]. In this chapter, as in most of the book, we constrain the unknowns (radiance r and depth map s) to live on a finite grid. However, this is not enough to guarantee that the solution is unique. To this end, one has to introduce additional knowledge or assumptions on the unknowns, such as the positivity of the radiance and the smoothness of the depth map. Notice that in doing so, one has also to make sure that a solution always exists. This leads to the following *regularization term* [3] added to the *data fidelity* term (5.1) yielding a composite energy functional which we call E

$$E(s, r) \doteq \Phi\left(J|I_s^r\right) + \alpha\|\nabla s\|^2. \tag{5.2}$$

Here α is a tuning parameter that controls the amount of regularity enforced. The problem of finding $\hat{s}$ and $\hat{r}$ from defocused images J can then be written as

$$\hat{s}, \hat{r} = \arg\min_{s, r} E(s, r) \qquad s > 0, \ r \geq 0. \tag{5.3}$$

In the rest of this chapter we develop a method to estimate both s and r by solving this optimization problem. This is our starting point in the same way equation (4.2) was the starting point for the least-squares solution. However, now we do not eliminate r and reduce the problem to finding the lone depth map. Rather, we develop iterative minimization techniques to solve the infinite-dimensional optimization problem directly.

5.2   Alternating minimization

Because there is no closed-form solution to (5.3), we concentrate on an iterative "gradient-like" scheme that, starting from some initial condition, reduces the cost

---

[2]If one is willing to give away the assumption of Poisson noise, then one can easily define a symmetric version of the I-divergence by summing or averaging $\Phi(I_s^r|J)$ and $\Phi(J|I_s^r)$.

[3]The reader may want to revisit Section 3.5 and if necessary Appendix F at this point.

function, with the hope of converging at least to a local minimum. There is naturally no guarantee that the solution will be the global minimum starting from any initial condition. There are several heuristics one can employ to improve the chances of converging to the global minimum. In Appendix E, Section E.4 we give the MATLAB® code of one such heuristic; alternatively, one can employ Monte Carlo methods that asymptotically guarantee global convergence [Liu, 2001], given enough patience. To set the stage, therefore, we assume that we have an initial estimate of r, which we call r_0. For instance, a natural choice of r_0 is one of the measured images. Then, we break down the problem (5.3) into the iterative minimization of the following two problems. Starting from $k = 0$, solve

$$\begin{cases} s_{k+1} \doteq \arg\min_s E(s, r_k) \\ r_{k+1} \doteq \arg\min_r E(s_{k+1}, r) \end{cases} \qquad (5.4)$$

until some stopping condition is satisfied. This clearly leads to a local minimum of E, because E decreases at each iteration, and it is bounded below by zero:

$$0 \leq E(s_{k+1}, r_{k+1}) \leq E(s_{k+1}, r_k) \leq E(s_k, r_k). \qquad (5.5)$$

However, solving the two optimization problems in equation (5.4) may be overkill. In order for the sequence $\{E(s_k, r_k)\}$ to converge it suffices that, at each step, we choose s_k and r_k in such a way as to decrease the value of E, so that equation (5.5) holds:

$$\begin{cases} s_{k+1} & E(s_{k+1}, r) \leq E(s_k, r) \qquad r = r_k \\ r_{k+1} & E(s, r_{k+1}) \leq E(s, r_k) \qquad s = s_{k+1}. \end{cases} \qquad (5.6)$$

The first step of the iteration, to update the depth map s_k, can be realized in a number of ways. We start by looking at an update of the surface in the direction of the gradient of the cost function E in (5.3), as usual in gradient descent minimization. In order to compute the gradient we must compute the derivative of E with respect to s, a function. This can be done using the techniques of calculus of variations (see Appendix B). For the benefit of the reader who is not familiar with these techniques, we report the detailed calculations in the following. The first order variation of E with respect to the depth map s in the direction $g_s : \Omega \to (0, \infty)$ is

$$\delta E = \sum_{i=1}^{K} \int_{\Omega} \left(1 - \frac{J(y, v_i)}{I_s^{r_{k+1}}(y, v_i)}\right) \int_{\Omega} \frac{\partial h_s^{v_i}}{\partial s}(y, x) r_{k+1}(x) dx g_s(y) dy$$
$$+ 2\alpha \int_{\Omega} \nabla s(y) \cdot \nabla g_s(y) dy.$$

$$(5.7)$$

By integrating by parts, and using the fact that $g_s(y) = 0, \forall y \in \partial\Omega$, this yields the gradient

$$\nabla_s E(s, r_{k+1})(y) = \sum_{i=1}^{K} \left(1 - \frac{J(y, v_i)}{I_s^{r_{k+1}}(y, v_i)}\right) \frac{\partial I_s^{r_{k+1}}}{\partial s}(y, v_i) - 2\alpha \triangle s(y); \quad (5.8)$$

where

$$\frac{\partial I_s^{r_{k+1}}}{\partial s}(y, v_i) = \int_{\Omega} \frac{\partial h_s^{v_i}}{\partial s}(y, x) r_{k+1}(x) dx. \tag{5.9}$$

The simplest gradient descent updates the surface s by performing a step proportional to β in the direction of the gradient as follows

$$s_{k+1} = s_k - \beta \nabla_s E(s_k, r_{k+1}) \tag{5.10}$$

with $\beta > 0$ a small scalar. Notice that the iteration stops whenever the gradient (5.8) is 0, that is, when the Euler–Lagrange equation is satisfied, a necessary condition for an extremum (see Appendix B). This iteration, natural as it may appear, has a number of shortcomings. The most important is that equation (5.8), and therefore (5.10), depends on the magnitude of the radiance. This means that dark regions move slower and light regions move faster, clearly not a desirable feat. We can correct this behavior by inserting into (5.10), at each iteration k, a positive-definite kernel M_k. To equalize the velocity of the depth map s it is then reasonable to consider the following choice,

$$M_k(y, \bar{y}) = \delta(y - \bar{y}) \left(\max \left\{ \epsilon_M, \sqrt{\sum_{i=1}^{K} \left(\frac{\partial I_s^{r_k}}{\partial s}(y, v_i) \right)^2} \right\} \right)^{-1}, \tag{5.11}$$

which is the best approximation available at iteration k of the magnitude of the variation of the images with respect to a variation of the shape, and where $\epsilon_M > 0$ is a small number to avoid division by zero. We call the kernel M_k a *preconditioner*. Notice that the preconditioner M_k in the equation above is positive-definite by construction. With this choice, the iteration (5.10) becomes

$$s_{k+1} = s_k - \beta \langle M_{k+1}, \nabla_s E(s_k, r_{k+1}) \rangle, \tag{5.12}$$

which does not depend on the magnitude of the radiance nor on the number of defocused images available. Clearly, other choices are possible as long as they have similar properties and convergence behavior. Notice that the preconditioner does not compromise the minimization of the cost functional, as the following calculation proves.

$$\begin{aligned} \frac{\partial E(s_t, r_{k+1})}{\partial t} &= \int_{\Omega} \nabla_s E(s_t, r_{k+1})(y) \frac{\partial s_t}{\partial t}(y) dy \\ &= -\int_{\Omega} \nabla_s E(s_t, r_{k+1})(y) \langle M_{k+1}, \nabla_s E(s_t, r_{k+1}) \rangle dy < 0 \end{aligned} \tag{5.13}$$

whenever $M_{k+1} > 0$ and $\|\nabla_s E(s_t, r_{k+1})\| \neq 0$, and where we have used the chain rule together with the following approximation

$$\frac{\partial s_t}{\partial t}(y) = -\langle M_{k+1}, \nabla_s E(s_t, r_{k+1}) \rangle. \tag{5.14}$$

Another very important shortcoming is the relative convergence rate between the data term and the regularization term. As we can see in equation (5.8), although the data term depends on the image intensities, the regularization term does not. Hence, the preconditioner M_{k+1} equalizes the convergence rate of the data term, but also unbalances the regularization term. Unfortunately, the strategy that we used above cannot be applied in a straightforward manner to the equalization of the regularization term. We need to modify the regularization term in the original cost functional so that the functional derivative results in

$$\nabla_s E(s, r_{k+1})(y) = \sum_{i=1}^{K} \left(1 - \frac{J(y, v_i)}{I_s^{r_{k+1}}(y, v_i)}\right) \frac{\partial I_s^{r_{k+1}}}{\partial s}(y, v_i) \\ - 2\alpha \langle M_{k+1}^{-1}, \triangle s \rangle \tag{5.15}$$

and then, after preconditioning, the gradient is

$$\langle M_{k+1}, \nabla_s E(s, r_{k+1}) \rangle = \langle M_{k+1}, \sum_{i=1}^{K} \left(1 - \frac{J(\cdot, v_i)}{I_s^{r_{k+1}}(\cdot, v_i)}\right) \frac{\partial I_s^{r_{k+1}}}{\partial s}(\cdot, v_i) \rangle \\ - 2\alpha \triangle s. \tag{5.16}$$

A first issue with obtaining equation (5.16) is that the preconditioner M_{k+1} depends on both the depth map s and the radiance r, and thus introduces spurious undesired terms in the gradient. To overcome this problem we consider using $M_\infty = \lim_{k \to \infty} M_k$ in the modified regularization term, instead of M_{k+1} and we use equation (5.16) to approach the exact gradient asymptotically because we do not know M_∞. The energy term that corresponds to such regularization might not even exist, and if it exists its computation might be nontrivial. However, the existence or the evaluation of such energy term is not necessary as long as we know where the preconditioned gradient iteration converges to. Indeed, by construction, the preconditioned gradient is zero only at the extrema of $\|\nabla s\|^2$, and has the same convergence behavior of the original gradient as we showed in equation (5.13). Hence, we can directly employ equation (5.16) in our iteration. A more thorough analysis of the preconditioned gradient flow would also require proving the existence and the uniqueness of the solution that the combined gradient flow yields. We do not report such proofs here as they are out of the scope of this book. The second step in this calculation is obtained from the Kuhn–Tucker conditions [Luenberger, 1968] associated with the problem of minimizing E for fixed s under positivity constraints for $\hat{r}$:

$$\sum_{i=1}^{K} \int_\Omega \frac{h_s^{v_i}(y, x) J(y, v_i)}{I_s^{\hat{r}}(y, v_i)} dy = \begin{cases} = \sum_{i=1}^{K} \int_\Omega h_s^{v_i}(y, x) dy & \forall x \mid \hat{r}(x) > 0 \\ \leq \sum_{i=1}^{K} \int_\Omega h_s^{v_i}(y, x) dy & \forall x \mid \hat{r}(x) = 0. \end{cases} \tag{5.17}$$

Because such conditions cannot be solved in closed form, we look for an iterative procedure for r_k that will converge to $\hat{r}$ as a fixed point. Snyder et al. [Snyder et al., 1992] have shown that the following function

$$F_s^{r_k}(x) \doteq \frac{1}{\sum_{i=1}^{K} \int_{\Omega} h_s^{v_i}(y,x)dy} \sum_{i=1}^{K} \int_{\Omega} \frac{h_s^{v_i}(y,x)J(y,v_i)}{I_s^{r_k}(y,v_i)} dy \qquad (5.18)$$

can be used to define an iteration that has the Kuhn–Tucker conditions as the fixed point:

$$r_{k+1}(x) = r_k(x)F_s^{r_k}(x). \qquad (5.19)$$

To see that this is the case, notice that equation (5.19) has a fixed point for $k \to \infty$ whenever $F_s^{r_\infty}(x) = 1$ and $r_\infty(x) > 0$, or $r_\infty(x) = 0$. The first case corresponds to the first of the Kuhn–Tucker conditions in equation (5.17). Instead, the case of x such that $r_\infty(x) = 0$ requires proving the continuity of $F_s^{r_k}$ with respect to r_k (see [Snyder et al., 1992]). Then, choose $r_0(x) > 0$ and assume by contradiction that $F_s^{r_\infty}(x) > 1$; by the continuity of $F_s^{r_k}(x)$ there exists a k^* such that $F_s^{r_k}(x) > 1$, $\forall k > k^*$ and hence equation (5.19) iterated for $k \to \infty$ yields the following contradiction:

$$0 = r_\infty(x) = r_0(x)\Pi_{k=0}^{\infty}F_s^{r_k}(x) \to \infty. \qquad (5.20)$$

We suggest the interested reader review [Snyder et al., 1992] for a more thorough analysis. What matters in the end is that, when the iteration (5.19) converges, we are guaranteed that the conditions (5.17) are satisfied. It is important to point out that this iteration decreases the functional E not only when we use the exact kernel h_s, but also with any other kernel satisfying the positivity constraints, for instance, the kernel based on the current estimate of s (which may still be far from the correct one) during the alternating minimization.

Proposition 5.1. *Let r_0 be a positive real-valued function, and let the sequence r_k be defined according to equation (5.19). Then $E(s, r_{k+1}) \leq E(s, r_k) \, \forall \, k > 0$ and for all admissible surfaces s. Furthermore, equality holds if and only if $r_{k+1} = r_k$.*

Proof. See Appendix C.4. □

Remark 5.2 (Minimization of E does not necessarily mean success). *Note that the fact that the sequence $E(s_k, r_k)$ converges to a minimum is no guarantee that s_k and r_k converge to the correct estimates of the shape and radiance of the scene. Even if the sequence $E(s_k, r_k)$ converges to 0, s_k and r_k could converge to something other than the true shape and radiance, or could not converge and wander around while keeping the value of $E = 0$. For instance, trivially, if we take images of a white scene, and set r to be uniformly white, E would be 0 for any admissible s. Convergence of $E(s_k, r_k)$ is only necessary, but not sufficient, for convergence of shape and radiance, and the fact that the residual $E(\hat{s}, \hat{r})$ is small is also no indication that the algorithm has yielded a good estimate. In order to conclude that the minimization of E yields the correct shape and radiance, the observability conditions discussed in Chapter 3 will also have to be satisfied. We* *therefore encourage the reader to revisit that chapter before moving on to the implementation of the iterative minimization algorithm.*

5.3   Implementation

The first step to implement the proposed algorithm is to choose a finite-dimensional representation for both the radiance r and the shape s. We define both unknowns on a fixed grid whose coordinates correspond to the sensor array. Such a choice is suggested by the fact that the highest resolution at which reconstruction is possible. [4] depends on the smallest blurring radius (in addition to the number of input images) Because the intensity value at each pixel is the (area) integral of all the photons hitting a single sensor element (e.g., a CCD), the smallest blurring radius is limited by the physical size of the CCD cell. Furthermore, diffraction and similar phenomena also contribute to increasing the minimum blurring radius. [5]
As mentioned before, we could initialize the radiance r_0 with one of the input images and the depth map s_0 with some approximate method (see Appendix E, Section E.2 for one such method). We start by initializing the depth map s_0 and then use it to determine at each pixel location the sharpest image. Then, we initialize the radiance by assigning the intensity of the sharpest image at each pixel as described in Table 5.1. This choice is guaranteed to be admissible because each image is strictly positive and its values are finite. Then, we alternate the iterations on the depth map s and radiance r as described in the previous section. The algorithm is summarized in Table 5.1 and the corresponding MATLAB® implementation is given in Appendix E, Section E.2.

5.4   Examples

To illustrate the algorithm above we show some tests on synthetic and real images. The estimated depth maps and radiances are shown together with the ground truth in the case of synthetic images, whereas in the case of real images validation is performed qualitatively by visual inspection.

5.4.1   Examples with synthetic images

In the synthetic data set we evaluate the algorithm on a number of different shapes: a wave, a slope, and a scene made of equifocal planes at different depths. In

---

[4] A simple upper bound of the highest accuracy achievable, can be obtained from a straightforward application of the sampling theorem [Oppenheim and Schafer, 1975].

[5] In our current implementation we allow the kernel h_s to be a Dirac delta. Hence, we consider the case where the radiance r is strictly positive, supported by the fact that the previous analysis is still valid under these assumptions.

Table 5.1. I-divergence algorithm to compute a solution of shape from defocus via the alternating minimization described in Section 5.2.

Algorithm (I-divergence)

1. Given: Calibration parameters (from knowledge of the camera; see Appendix D for a simple calibration procedure) $\{v_i\}_{i=1,\ldots,K}$, F, D, γ, images $J(y, v_1), \ldots, J(y, v_K)$, a chosen threshold $\epsilon > 0$, a step size β, a value ϵ_M, and a regularization coefficient α,

2. Initialize shape s_0 by using the algorithm in Appendix E.4; initialize the radiance r_0 by letting

$$r_0(y) = J(y, v_i) \quad \text{where } i = \min_n v_n \left| \frac{1}{F} - \frac{1}{v_n} - \frac{1}{s_0(y)} \right|,$$

3. While $E(s_k, r_k) - E(s_{k+1}, r_{k+1}) > \epsilon$

   - Perform iteration (5.19);
   - Perform iteration (5.12).

addition, to illustrate the insensitivity of the algorithm to the magnitude of the radiance, we choose radiances with three very different intensity levels. In the data set of real images the equifocal planes are at 520 mm and 850 mm, with focal length 12 mm, F-number 2, and $\gamma = 10^4$ pixel^2 / mm^2. In Figures 5.1, 5.3, and 5.5 we show experiments conducted on a wave, a slope, and piecewise-constant depth map, respectively. In each figure we show the input images on the top row, and the true radiance on the bottom-left image; on the bottom-right we show the estimated radiance. In Figures 5.2, 5.4 and 5.6 we display the true depth map as a grayscale intensity map (light intensities correspond to points close to the camera, and dark intensities correspond to points far from the camera) on the top-left image and as a mesh on the bottom-left image. Similarly, the estimated depth map is displayed as a grayscale intensity map on the top-right image and as a mesh on the bottom-right image. Notice that the estimated depth map converges uniformly towards the true depth map in all cases, despite the different intensity (top to bottom) and the different level of sharpness (left to right) of the true radiance (see, for instance, the bottom-left image in Figure 5.1).

Because the ground truth is available to us, we can also compute the discrepancy between the true depth map and radiance and the estimated ones. We compute two types of discrepancies:

$$
\begin{aligned}
Err_0 &= \sqrt{E\left[\left(\xi - \hat{\xi}\right)^2\right]} \\
Err_1 &= \sqrt{E\left[\left(\frac{\hat{\xi}}{\xi} - 1\right)^2\right]},
\end{aligned}
\tag{5.21}
$$

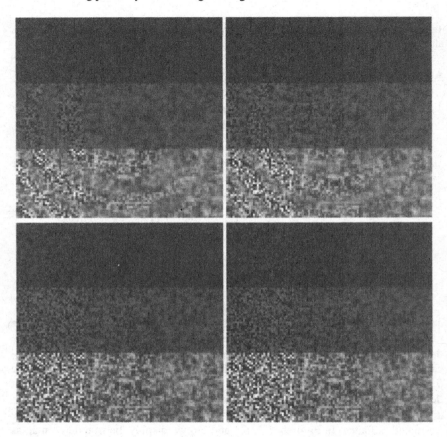

Figure 5.1. Wave data set. Top-left: Near-focused original image. Top-right: Far-focused original image. Bottom-left: Original radiance. Bottom-right: Estimated radiance.

where $E[\cdot]$ denotes the average over [6] Ω, ξ is the ground truth (the radiance or the depth map), and $\hat{\xi}$ is the corresponding estimate. The first one, Err_0, measures the relative error in meters in the case of the depth map, and in grayscale values (0 to 256 and denoted by u) in the case of the radiance; the second, Err_1, measures the absolute relative error and it is therefore unitless. We show the absolute and relative errors for each of the synthetic experiments in Table 5.2.

5.4.2   Examples with real images

In the case of real data we have the same camera settings as in the case of the synthetic data (they were chosen so to facilitate comparison) apart from γ that is

---

[6]In the experiments, the average does not include the boundary of Ω, $\partial\Omega$, which is three pixels wide.

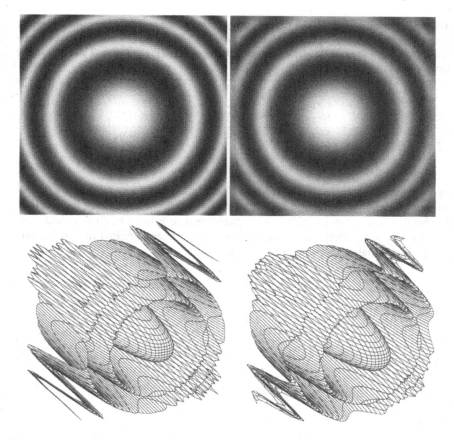

Figure 5.2. Wave data set. Top-left: Original depth map in grayscale intensities. Top-right: Estimated depth map in grayscale intensities. Bottom-left: Original depth map displayed as a mesh. Bottom-right: Estimated depth map displayed as a mesh.

set to $1.5 \cdot 10^4$ pixel^2 / mm^2. The scene is composed of a box at the bottom and a cylinder and a small box on the top (see top row in Figure 5.7). We capture the two defocused images shown in the top row in Figure 5.7 and obtain the radiance shown in the bottom-left image and the depth map in the bottom-right image. The depth map is also displayed as a mesh and viewed from different vantage points in Figure 5.8 to better appreciate the results.

5.5   Summary

When the discrepancy between the ideal image and the measured one is modeled as a Poisson process, shape and radiance can be estimated by minimizing the I-divergence between measured and ideal images. In this case, we do not separate

Table 5.2. Absolute and relative discrepancies between ground truth and estimated depth maps and radiances. The absolute discrepancies, Err_0, are in meters (m) in the case of the depth map and in grayscale intensities (u) with values ranging between 0 and 256 in the case of the radiances.

		Wave	Slope	Boxes
Depth map	Err_0	0.012 m	0.001 m	0.018 m
	Err_1	1.7%	0.08%	1.15%
Radiance	Err_0	0.165 u	0.010 u	0.164 u
	Err_1	0.6%	0.4%	1.3%

the problem of shape from defocus from that of image restoration. Instead, we derive an iterative minimization algorithm that solves for both, at least in principle. In practice, convergence can only be guaranteed to a local minimum, and even when the cost functional is minimized there is no guarantee that the minimizers (shape and radiance) are the correct ones.

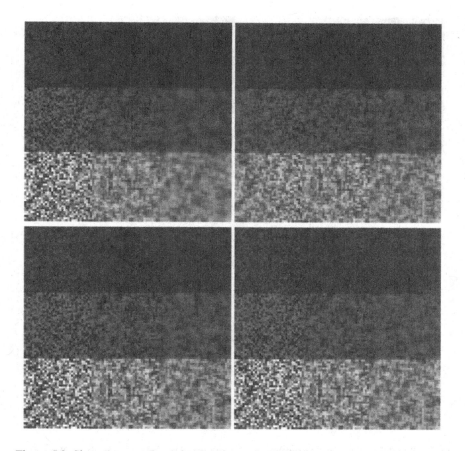

Figure 5.3. Slope data set. Top-left: Near-focused original image. Top-right: Far-focused original image. Bottom-left: Original radiance. Bottom-right: Estimated radiance.

Figure 5.4. Slope data set. Top-left: Original depth map in grayscale intensities. Top-right: Estimated depth map in grayscale intensities. Bottom-left: Original depth map displayed as a mesh. Bottom-right: Estimated depth map displayed as a mesh.

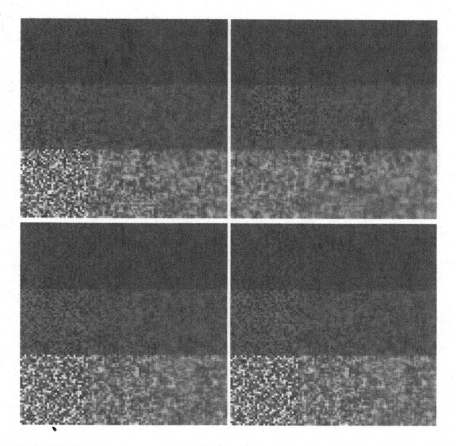

Figure 5.5. Piecewise-constant data set. Top-left: Near-focused original image. Top-right: Far-focused original image. Bottom-left: Original radiance. Bottom-right: Estimated radiance.

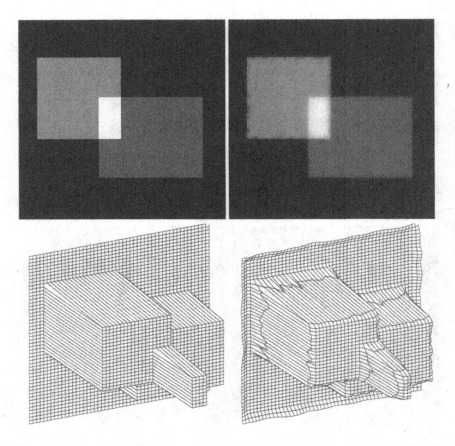

Figure 5.6. Piecewise-constant data set. Top-left: Original depth map in grayscale intensities. Top-right: Estimated depth map in grayscale intensities. Bottom-left: Original depth map displayed as a mesh. Bottom-right: Estimated depth map displayed as a mesh.

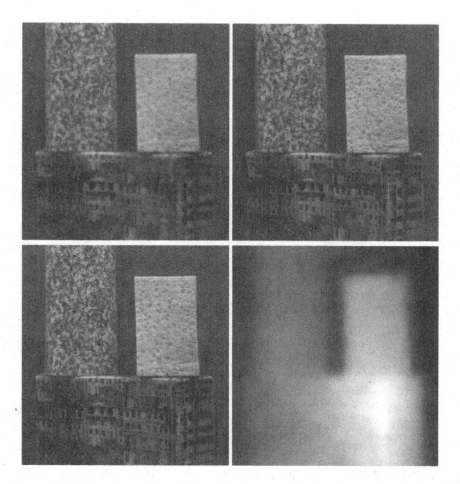

Figure 5.7. Real data set. Top-left: Far-focused original image. Top-right: Near-focused original image. Bottom-left: Estimated radiance. Bottom-right: Estimated depth map.

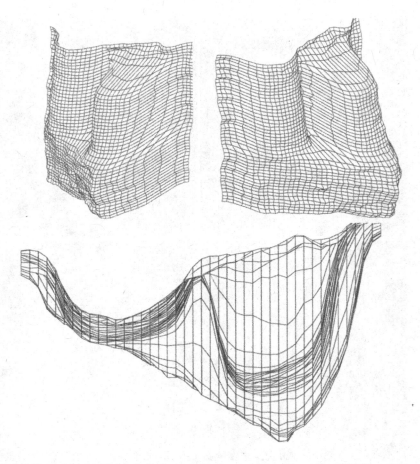

Figure 5.8. Real data set. Top-left: Estimated depth map viewed from the right. Top-right: Estimated depth map viewed from the left. Bottom: Estimated depth map viewed from the top.

6

Defocus via diffusion: Modeling and reconstruction

In Chapter 2 we have described the formation of an image through a thin lens as the integral of the radiance of the scene, that can be thought of as the "perfect" (or sharp, or deblurred) image, against a kernel that depends on the geometry of the imaging device (equation (2.9)). In other words, a measured image is just a blurred version of the radiance, or energy distribution, of the scene. There, we thought of the radiance of the scene as describing the heat distribution on a metal plate, with the temperature representing the intensity at a given point, and heat diffusion on the plate blurring the original heat distribution (Figure 2.6). This process can be simulated by solving the *heat equation*, a particular kind of partial differential equation (PDE), with "time" being an index that describes the amount of blur.

In this interpretation, our problem entails "reversing" the heat equation to recover the geometry of the scene (encapsulated in the diffusion coefficient) and its radiance, the deblurred image, from blurred ones. Unfortunately, this requires reversing a diffusion, which is an ill-posed problem. However, we show how this problem can be solved without time reversals, using only forward diffusion.

The basic idea consists of bypassing the recovery of the original, deblurred, radiance by introducing the notion of *relative blur*. Given two images, at every spatial location one of the two will be blurrier than the other, depending on the structure of the scene (an image can be sharper than the other at one place, but blurrier at another place, as in Figure 6.1). One can then diffuse locally the sharper image towards the blurrier one, and the amount of diffusion required to match the two can be related to the structure of the scene. We describe this process in detail in this chapter. The extension of these ideas to multiple images is not entirely straightforward, and is therefore addressed in Section 6.6.

6.1   Blurring via diffusion

In Section 2.4 we introduced a model for generating defocused images via a diffusion process, described by the heat equation (2.30). Such a model is based on the relationship between defocus and diffusion, which we recall as follows. We first assume, for simplicity, that the point-spread function (PSF) can be approximated by a shift-invariant Gaussian function:

$$h(y, x) = \frac{1}{2\pi\sigma^2} e^{-\|x-y\|^2/(2\sigma^2)}, \tag{6.1}$$

where the blurring parameter σ is given by

$$\sigma = \gamma b \tag{6.2}$$

for a certain calibration coefficient γ. The blurring parameter is related to the depth map s via the blur radius b (see Section 2.4) so that

$$b = \frac{Dv}{2} \left| \frac{1}{F} - \frac{1}{v} - \frac{1}{s} \right| \tag{6.3}$$

and v is the focus setting of the camera. As we discussed in Section 2.4, the blurred image $I(y)$, expressed as an integral in (2.9), can be obtained as the solution $u(y, t)$ of the heat equation (2.30), which we repeat here for convenience, using y as the argument because the image $I(y)$ is obtained as the solution $u(y, t)$ at a particular "time" t:

$$\begin{cases} \dot{u}(y, t) = c\triangle u(y, t) & c \in [0, \infty) \quad t \in (0, \infty) \\ u(y, 0) = r(y) & \forall\, y \in \mathbb{R}^2 \end{cases}, \tag{6.4}$$

where, again, r is the radiance of the scene, assumed zero outside of Ω. Recall that $\triangle$ denotes the Laplacian operator (we defined it in Section 2.4). [1]

Here t does not indicate physical time, as it does in the context of heat or chemical diffusion. Rather, it is an independent variable that regulates the total "amount of diffusion" that the initial condition $r(y)$ has undergone. Therefore, at a particular time $t = t_1$, we obtain an image $J(y) = u(y, t_1), \forall y \in \Omega \subset \mathbb{R}^2$ from the radiance $r(y)$; the reader who is still confused may want to revisit Section 2.4 at this point. The blurring parameter σ is now related to the diffusion coefficient c via equation (2.32), that is,

$$\sigma^2 = 2tc. \tag{6.5}$$

The ambiguity between the time index t and the diffusion coefficient c (they appear as a product in σ) is fixed later by choosing t to be a prescribed constant.

---

[1] It takes sharp eyes to distinguish it from $\triangle$, the difference operator, but the symbols are indeed different ($\triangle$ versus $\triangle$) and the reader should be able to tell them apart easily from the context.

6.2 Relative blur and diffusion

The image $J(y)$ is obtained from the diffusion equation (6.4) starting from the radiance $r(y)$, which we do not know. Rather than having to estimate it from two or more images, in this section we show how to arrive at a model of the relative blur between two images that does not depend on the radiance. Suppose we collect two images $J_1(y) \doteq J(y, v_1)$ and $J_2(y) \doteq J(y, v_2)$ for two different focus settings v_1 and v_2 corresponding to blurring parameters σ_1 and σ_2. Also, to keep the presentation simple, suppose that $\sigma_1 < \sigma_2$ (i.e., image J_1 is more focused than image J_2 at every pixel); we remove this assumption later. Then, we can write image J_2 by substituting the expression of r in terms of image J_1 as follows.

$$
\begin{aligned}
J_2(y) &= \int \frac{1}{2\pi\sigma_2^2} e^{-\|x-y\|^2/(2\sigma_2^2)} r(x) dx \\
&= \int \frac{1}{2\pi(\sigma_2^2 - \sigma_1^2)} e^{-\|x-y\|^2/(2\sigma_2^2 - 2\sigma_1^2)} \\
&\qquad \int \frac{1}{2\pi\sigma_1^2} e^{-\|\bar{x}-x\|^2/(2\sigma_1^2)} r(\bar{x}) d\bar{x} dx \\
&= \int \frac{1}{2\pi\Delta\sigma^2} e^{-\frac{\|x-y\|^2}{2\Delta\sigma^2}} J_1(x) dx,
\end{aligned}
\tag{6.6}
$$

where $\Delta\sigma^2 \doteq \sigma_2^2 - \sigma_1^2$ and $\Delta\sigma$ is called the *relative blurring* between image J_1 and image J_2. Now, J_2 in equation (6.6) can also be interpreted as a solution of the heat equation (6.4), but with J_1, rather than r, as an initial condition. Let t_1 and t_2 be the two time instants computed from equation (6.5) with a fixed diffusion coefficient c for blurring parameters σ_1 and σ_2, respectively; then, the solution of equation (6.4) satisfies $u(y, t_1) = J_1(y)$ and $u(y, t_2) = J_2(y)$, $\forall y \in \Omega$, and we can write

$$
\begin{cases}
\dot{u}(y, t) = c\Delta u(y, t) & c \in [0, \infty) \\
u(y, t_1) = J_1(y) & \forall y \in \Omega.
\end{cases}
\tag{6.7}
$$

If $\sigma_1 > \sigma_2$, one can simply switch J_1 with J_2.

Equation (6.7) models the *relative diffusion* between image J_1 and image J_2. This allows us to eliminate r as an unknown, and concentrate our resources on recovering only the geometry of the scene.

For simplicity of notation, rather than using the solution $u(\cdot, t)$, we can consider the time-shifted version $u(\cdot, t - t_1)$ (or set $t_1 = 0$) so that

$$
\begin{cases}
\dot{u}(y, t) = c\Delta u(y, t) & c \in [0, \infty) \\
u(y, 0) = J_1(y) & \forall y \in \Omega
\end{cases}
\tag{6.8}
$$

with $u(y, \Delta t) = J_2(y)$, where Δt is defined by the equation

$$
\Delta\sigma^2 = 2(t_2 - t_1)c \doteq 2\Delta t c.
\tag{6.9}
$$

One can view the time Δt as the variable encoding the change in focus settings, and the diffusion coefficient c as the variable encoding the depth map s via

$$c = \frac{\Delta \sigma^2}{2\Delta t} = \frac{\gamma^2 \left(b_2^2 - b_1^2\right)}{2\Delta t}, \tag{6.10}$$

where for $i = 1, 2$

$$b_i = \frac{Dv_i}{2} \left| \frac{1}{F} - \frac{1}{v_i} - \frac{1}{s} \right|. \tag{6.11}$$

This discussion holds as long as the PSF is shift-invariant. As we have seen in Section 2.4, this corresponds to the entire image being blurred uniformly, or "isotropically," at all locations, which occurs only if the scene is flat and parallel to the image plane (i.e., it is equifocal). Naturally, such scenes are rather uninteresting; therefore, to handle non flat scenes, we need to extend the diffusion analogy to equations that allow for a space-varying blur.

6.3   Extension to space-varying relative diffusion

As we have seen in Section 2.4.2, when the surface s is not an equifocal plane, the corresponding PSF is shift-varying and we cannot use the homogeneous heat equation to model defocused images. Therefore, we have introduced the inhomogeneous diffusion equation (2.34) by allowing the coefficient c to vary spatially as a function of the location on the image. Similarly, in the context of relative diffusion, we can extend equation (6.8) to

$$\begin{cases} \dot{u}(y,t) = \nabla \cdot (c(y)\nabla u(y,t)) & t \in (0, \infty) \\ u(y,0) = J_1(y) & \forall y \in \Omega \end{cases}, \tag{6.12}$$

where $u(y, \Delta t) = J_2(y)$, $\forall y \in \Omega$. Recall that in Section 2.4.2 we required the diffusion coefficient $c : \Omega \mapsto \mathbb{R}$ which satisfies a number of properties in order to guarantee that energy is conserved in the imaging process (2.29). We still assume that J_1 is more focused than J_2 (i.e., $\sigma_1 < \sigma_2$), for simplicity.

Remark 6.1. *There are other space-varying models that one could use in place of (6.12), for instance, the oriented-Laplacian [Tschumperle and Deriche, 2003]. Each has its own advantages, but none is exactly equivalent to the space-varying convolutional model derived in Section 2.2 or to more accurate models derived from diffraction theory [Hopkins, 1955]. Furthermore, the point-spread function of real optical systems can deviate substantially from the simple models that we describe here due to various sources of aberration, including parting from linearity. Hence, the use of one model or another may be decided based on precise knowledge of the imaging device or the specific application at hand or on computational resources. In this book, for the sake of clarity and simplicity, we focus only on model (6.12) as a prototype.*

Figure 6.1. Two defocused images J_1 and J_2. The apple in the image on the left (J_1) is sharper than the apple in the image on the right (J_2). At the same time, the background of the image on the right is sharper than the corresponding background of the image on the left.

6.4 Enforcing forward diffusion

In the derivation of equation (6.12) we made the assumption that J_1 is more focused than J_2. This assumption is necessary in order to guarantee that $c(y) \geq 0$, $\forall y \in \Omega$, and hence that (6.12) involves only numerically stable and well-behaved forward diffusion, and not backward diffusion. However, as illustrated in Figure 6.1, this assumption is in general not valid. The apple in image J_1 (left) is more focused than the corresponding apple in image J_2 (right). On the other hand, the background is more focused in image J_2. As a consequence, the relative blurring $\Delta\sigma$ between J_1 and J_2, as well as the corresponding diffusion coefficient c, may be either positive or negative, depending on at which location of the image domain Ω they are evaluated. In this case, model (6.12) involves both backward diffusion and forward diffusion. Backward diffusion is numerically unstable and has the undesirable effect of amplifying high-frequency content as time increases.

A simple way to avoid backward diffusion in the imaging model (6.12) is to restrict the solution u to the set Ω_+ where the diffusion coefficient is positive, that is,

$$\Omega_+ \doteq \{y \in \Omega \mid c(y) > 0\} \tag{6.13}$$

and $u(y, \Delta t) = J_2(y)$, $\forall y \in \Omega_+$. On the remaining portion

$$\Omega_- \doteq \Omega \setminus \Omega_+ \doteq \{y \in \Omega \mid c(y) \leq 0\} \tag{6.14}$$

the diffusion coefficient c is negative[2] and J_2 is sharper than J_1. Hence, on Ω_- we can employ the following model, that simply switches J_1 and J_2

$$\begin{cases} \dot{u}(y, t) = \nabla \cdot (-c(y)\nabla u(y, t)) & t \in (0, \infty) \\ u(y, 0) = J_2(y) & \forall y \in \Omega_- \end{cases} \tag{6.15}$$

---

[2]We always consider the diffusion coefficient c to be computed via equation (6.9), where $\Delta\sigma$ is the relative blur between J_1 and J_2.

Figure 6.2. Visualization of the partition $\{\Omega_+, \Omega_-\}$ obtained from the images in Figure 6.1. The set Ω_+ is visualized by brightening the intensity of the original images, whereas the set Ω_- is visualized by darkening the intensity of the original images.

with $u(y, \Delta t) = J_1(y)$, $\forall y \in \Omega_-$. In Figure 6.2 we visualize the partition $\{\Omega_+, \Omega_-\}$ obtained from the images in Figure 6.1. The set Ω_+ is visualized by brightening the original images, whereas the set Ω_- is visualized by darkening them.

By using the partition $\{\Omega_+, \Omega_-\}$ we obtain an imaging model for pairs of defocused images J_1 and J_2 that involves only forward diffusion. This, however, requires knowledge of the partition $\{\Omega_+, \Omega_-\}$ that is not available. However, equation (6.13) tells us that the partition depends on the diffusion c which, in turn, depends on the unknown depth map s via equations (6.10) and (6.11). Therefore, we estimate the depth map along with recovering the partition.

6.5   Depth-map estimation algorithm

In previous sections we derived an idealized image-formation model in terms of the diffusion equation (6.4). First we eliminated the radiance from the model by introducing the notion of relative blur (Section 6.2); then, we extended the model to capture non planar scenes (Section 6.3) and, finally, we enforced forward diffusion by partitioning the image domain (Section 6.4). The result of these steps is an imaging model composed of equations (6.7) and (6.15), which we rewrite here in a more complete form, including boundary conditions, as

$$
\begin{cases}
\dot{u}(y, t) = \begin{cases} \nabla \cdot (c(y)\nabla u(y, t)), & y \in \Omega_+ \\ \nabla \cdot (-c(y)\nabla u(y, t)), & y \in \Omega_- \end{cases} & t \in (0, \infty) \\
u(y, 0) = \begin{cases} J_1(y), & \forall y \in \Omega_+ \\ J_2(y), & \forall y \in \Omega_- \end{cases} \\
0 = c(y)\nabla u(y, t) \cdot n(y), & \forall y \in \partial\Omega_+ \equiv \partial\Omega_+ \\
u(y, \Delta t) = \begin{cases} J_2(y), & \forall y \in \Omega_+ \\ J_1(y), & \forall y \in \Omega_- \end{cases}
\end{cases}
\quad , \quad (6.16)
$$

where n is the unit vector orthogonal to $\partial\Omega_+$, the boundary of Ω_+, which coincides with the boundary of Ω_-. The boundary conditions of the diffusion equations are satisfied because, by construction, the diffusion coefficient $c(y) = 0$ for any $y \in \partial\Omega_+$ and $y \in \partial\Omega_-$. Also, recall that the diffusion coefficient c depends on the depth map s via equations (6.10) and (6.11), which can be explicitly written as

$$c(y) = \frac{\gamma^2 D^2}{8\Delta t} \left(v_2^2 \left(\frac{1}{F} - \frac{1}{v_2} - \frac{1}{s(y)} \right)^2 - v_1^2 \left(\frac{1}{F} - \frac{1}{v_1} - \frac{1}{s(y)} \right)^2 \right). \quad (6.17)$$

Notice that $c(y) = 0$ when

$$s(y) = \frac{(v_1 + v_2)F}{v_1 + v_2 - 2F} \quad (6.18)$$

or

$$s(y) = F \quad (6.19)$$

so that

$$\partial\Omega_-, \partial\Omega_+ = \left\{ y \;\middle|\; s(y) = \frac{(v_1 + v_2)F}{v_1 + v_2 - 2F} \text{ or } s(y) = F \right\} \quad (6.20)$$

and similarly, by direct substitution,

$$\begin{aligned} \Omega_+ &= \left\{ y \;\middle|\; 0 < s(y) < F \text{ or } s(y) > \frac{(v_1 + v_2)F}{v_1 + v_2 - 2F} \right\} \\ \Omega_- &= \left\{ y \;\middle|\; F < s(y) < \frac{(v_1 + v_2)F}{v_1 + v_2 - 2F} \right\}. \end{aligned} \quad (6.21)$$

Note that the sets Ω_+ and Ω_- depend on the 3-D structure of the scene; the depth s satisfying equations (6.18) and (6.19) corresponds to regions where both images undergo the same amount of defocus. Hence, if we were given a depth map $\tilde{s}$ and the calibration parameters of the camera, we could compute the diffusion coefficient c and the partition $\{\Omega_+, \Omega_-\}$ and then simulate equations (6.16). If the depth map $\tilde{s}$ coincides with the depth map s of the scene, then the solution of equations (6.16) must satisfy the condition $u(y, \Delta t) = J_2(y), \forall y \in \Omega_+$ and $u(y, \Delta t) = J_1(y), \forall y \in \Omega_-$.

This naturally suggests an iterative procedure to tackle the inverse problem of reconstructing shape from defocused images. Starting from an initial estimate of the depth map and the resulting partition (e.g., a flat plane), diffuse each image in the region where it is more focused than the other image, until the two become equal. The amount of diffusion required to match the two images encodes information on the depth of the scene. More formally, we can pose the problem as the minimization of the following functional,

$$\hat{s} = \arg\min_s \int_{\Omega_+} (u(y, \Delta t) - J_2(y))^2 dy + \int_{\Omega_-} (u(y, \Delta t) - J_1(y))^2 dy + \alpha\|\nabla s\|^2.$$

$$(6.22)$$

Here the first two terms take into account the discrepancy between the simulated image u and the measured images J_1 and J_2, and the third term imposes a smoothness constraint on the depth map s that is regulated by the parameter $\alpha > 0$. Notice that during the minimization of (6.22) both the partition and the depth map are estimated simultaneously. This is even more evident in the next section where we present a computational scheme to carry out the cost functional minimization.

6.5.1   Minimization of the cost functional

In this section we describe an algorithm to minimize the cost functional (6.22). The complete derivation of the algorithm is fairly lengthy. However, those who are not familiar with variational calculations will benefit from going through all the steps of the derivation. The reader will find all the calculations in Appendix B, along with a MATLAB® implementation in Appendix E. For the benefit of those who are familiar with these calculations, and to not disrupt the flow, we present the final solution here, but we urge others to pause at this point and familiarize themselves with Appendix B.2.1.

To simplify the notation, let

$$
\begin{aligned}
E_1(s) &\doteq \int H(c(y))|u(y, \Delta t) - J_2(y)|^2 \\
E_2(s) &\doteq \int H(-c(y))|u(y, \Delta t) - J_1(y)|^2 \\
E_3(s) &\doteq \alpha \|\nabla s\|^2,
\end{aligned}
\tag{6.23}
$$

where H denotes the Heaviside function (Section 2.3, equation (2.22)). Also, let

$$
E(s) \doteq E_1(s) + E_2(s) + E_3(s)
\tag{6.24}
$$

so that equation (6.22) can be rewritten as

$$
\hat{s} = \arg \min_s E(s).
\tag{6.25}
$$

To minimize the above cost functional we employ an iterative descent algorithm. This requires computing a sequence of depth maps s, indexed by an iteration index k, that converges to a local minimum of the cost functional, that is, a sequence $\{\hat{s}(y, k)\}_{k=1,2,...}$, such that $\hat{s}(y) = \lim_{k \to \infty} \hat{s}(y, k)$, $\forall y \in \Omega$ where $\hat{s}(y)$ is at least a local extremum of (6.25). At each iteration we update the depth map by moving along the direction opposite to the gradient of the cost functional with respect to the depth map. In other words, we let

$$
\frac{\partial \hat{s}(y, k)}{\partial k} \doteq -\nabla_s E(y),
\tag{6.26}
$$

where $\nabla_s E$ is the gradient of E with respect to the depth map s. We have shown in Section 5.2, equation (5.13), that the iteration thus defined decreases the cost functional as k increases. Using equation (6.24) we can split the computation of $\nabla_s E$ into the computation of $\nabla_s E_1$, $\nabla_s E_2$, and $\nabla_s E_3$. By the chain rule, the first term yields

$$\nabla_s E_1(y) = \nabla_c E_1(y) \frac{\partial c}{\partial s}(y), \tag{6.27}$$

where

$$\frac{\partial c}{\partial s}(y) = \frac{\gamma^2 D^2 (v_2 - v_1)}{4s^2(y)\Delta t} \left[(v_2 + v_1) \left(\frac{1}{F} - \frac{1}{s(y)} \right) - 1 \right] \tag{6.28}$$

and

$$\begin{aligned} \nabla_c E_1(y) &= -2H(c(y)) \int_0^{\Delta t} \nabla u(y,t) \cdot \nabla w_1(y, \Delta t - t) \, dt \\ &+ \delta(c(y)) \left(u(y, \Delta t) - J_2(y) \right)^2, \end{aligned} \tag{6.29}$$

where $w_1 : \Omega_+ \times [0, \infty) \mapsto \mathbb{R}$ satisfies the following (adjoint parabolic) equation (see Appendix B.2.1),

$$\begin{cases} \dot{w}_1(y,t) = \nabla \cdot (c(y)\nabla w_1(y,t)) & t \in (0, \infty) \\ w_1(y,0) = u(y, \Delta t) - J_2(y) & \\ c(y)\nabla w_1(y,t) \cdot n(y) = 0 & \forall y \in \partial\Omega_+ \end{cases} \tag{6.30}$$

Similarly, the second term in equation (6.24) yields

$$\nabla_s E_2(y) = \nabla_c E_2(y) \frac{\partial c}{\partial s}(y) \tag{6.31}$$

and

$$\begin{aligned} \nabla_c E_2(y) &= 2H(-c(y)) \int_0^{\Delta t} \nabla u(y,t) \cdot \nabla w_2(y, \Delta t - t) \, dt \\ &- \delta(c(y)) \left(u(y, \Delta t) - J_1(y) \right)^2, \end{aligned} \tag{6.32}$$

where $w_2 : \Omega_- \times [0, \infty) \mapsto \mathbb{R}$ satisfies the following (adjoint parabolic) equation (see Appendix B.2.1),

$$\begin{cases} \dot{w}_2(y,t) = \nabla \cdot (-c(y)\nabla w_2(y,t)) & t \in (0, \infty) \\ w_2(y,0) = u(y, \Delta t) - J_1(y) & \\ c(y)\nabla w_2(y,t) \cdot n(y) = 0 & \forall y \in \partial\Omega_- \end{cases} \tag{6.33}$$

Finally, the last term of equation (6.24) yields

$$\nabla_c E_3(y) = \alpha \Delta s(y). \tag{6.34}$$

We summarize the algorithm in Table 6.1. Again, we encourage those who are not familiar with these calculations to pause and consult Appendix B and Appendix E that are meant to facilitate an intuitive understanding of this material.

6.6   On the extension to multiple images

Extending the algorithm above to handle multiple images while avoiding restoration of the radiance results in a combinatorial problem of factorial complexity. In

Table 6.1. Summary of the depth map reconstruction algorithm via relative diffusion.

Algorithm (relative diffusion)

1. Given: Calibration parameters (from knowledge of the camera; see Appendix D for a simple calibration procedure) v_1, v_2, F, D, γ, two images J_1, J_2, a chosen threshold ϵ, regularization parameter α, and step size β, seek for the depth map $\hat{s}$ as follows.

2. Initialize depth map $\hat{s}(y, 0)$ with a plane at depth

$$\hat{s}(y, 0) = \frac{(v_1 + v_2)F}{v_1 + v_2 - 2F},$$

3. Compute the diffusion coefficient c and the partition $\{\Omega_+, \Omega_-\}$ via equations (6.17) and (6.21),

4. Simulate (i.e., numerically integrate) equations (6.7) and (6.15),

5. Using the solutions obtained at the previous step, simulate equations (6.30) and (6.33),

6. Compute the gradient of u and w and evaluate equations (6.29), (6.32), and (6.34),

7. Update the depth map $\hat{s}(y, k)$ with $\hat{s}(y, k+1) = \hat{s}(y, k) - \beta \nabla_s E(y)$,

8. Return to Step 3 until norm of gradient is below the chosen threshold ϵ.

practice, we have to consider the diffusion equations between all possible orderings of the images. To get a more concrete understanding of what happens with $K > 2$ images, consider the problem of "sorting" the images in increasing order of blur at each pixel, so that we are always applying forward diffusions to the input images. This leads to $K!$ possible orderings. In the case of two images, this amounts to simply determining two possible orderings, that can be easily tackled as we suggested in the previous section. However, to compute the solution in the general case, we need to simulate $K!$ different PDEs, one for each ordering (again, notice that in the case $K = 2$ we only need to simulate $K! = 2$ PDEs). Hence, both the number of equations we need to solve and the complexity of the algorithm depend on $K!$, which grows rather quickly as $K > 2$.

6.7   Examples

As we did in the previous chapter, to illustrate the algorithm above and validate it empirically, we test it on a number of synthetic and real images. The estimated depth maps are shown together with the ground truth in the case of synthetic images, whereas in the case of real images validation is performed qualitatively by visual inspection.

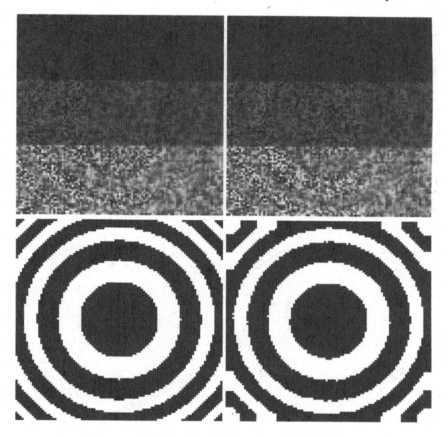

Figure 6.3. Wave data set. Top-left: Near-focused original image. Top-right: Far-focused original image. Bottom-left: True region Ω_+. Bottom-right: Estimated region Ω_+.

6.7.1   Examples with synthetic images

In the synthetic data set we evaluate the algorithm on a number of different shapes: a wave, a slope, and a scene made of equifocal planes at different depths. In addition, to illustrate the insensitivity of the algorithm to the magnitude of the radiance, we choose radiances with three different intensity levels. In the data set of real images the equifocal planes are at 520 mm and 850 mm, with focal length 12 mm, F-number 2 and $\gamma = 1.5 \cdot 10^4$ pixel^2 / mm^2. In Figures 6.3, 6.5, and 6.7 we show experiments conducted on a wave, a slope, and a piecewise-constant depth map, respectively. In each figure we show the input images on the top row, and the true region Ω_+ on the bottom-left image; on the bottom-right we show the estimated region Ω_+. In Figures 6.4, 6.6, and 6.8 we display the true depth as a grayscale intensity map (light intensities correspond to points close to the camera, and dark intensities correspond to points far from the camera) on the top-left image and as a mesh on the bottom-left image. Similarly, the estimated depth

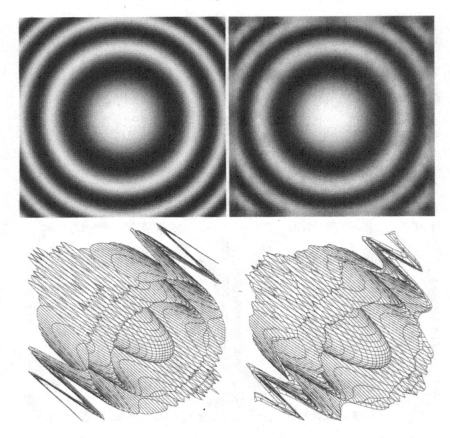

Figure 6.4. Wave data set. Top-left: Original depth map in grayscale intensities. Top-right: Estimated depth map in grayscale intensities. Bottom-left: Original depth map displayed as a mesh. Bottom-right: Estimated depth map displayed as a mesh.

is displayed as a grayscale intensity map on the top-right image and as a mesh on the bottom-right image. Notice that the estimated depth map converges evenly towards the true depth map in all cases, despite the different intensity (top to bottom) and the different level of sharpness (left to right) of the input images (see, for instance, the images on the top row of Figure 6.3).

Because the ground truth is made available to us, we can also compute the discrepancy between the true and the estimated depth map. As we did in the previous

Table 6.2. Absolute and relative discrepancies between ground truth and estimated depth maps. The absolute discrepancies, Err_0, are in meters (m).

		Wave	Slope	Boxes
Depth map	Err_0	0.016 m	0.0009 m	0.015 m
	Err_1	2.2%	0.13%	2.3%

chapter, we compute two types of discrepancies:

$$Err_0 = \sqrt{E\left[\left(\xi - \hat{\xi}\right)^2\right]}$$

$$Err_1 = \sqrt{E\left[\left(\frac{\hat{\xi}}{\xi} - 1\right)^2\right]}, \quad (6.35)$$

where ξ is the true depth map and $\hat{\xi}$ is the corresponding estimated depth map. $E[\cdot]$ denotes the average over [3] Ω. The first one, Err_0, measures the relative error in meters (m) in the case of the depth map; the second, Err_1, measures the absolute relative error and it is therefore unitless. We show the absolute and relative errors for each of the synthetic experiments in Table 6.2.

6.7.2 Examples with real images

In the case of real data we have the same camera settings as in the case of the synthetic data apart from γ that is set to $3 \cdot 10^4$ pixel^2 / mm^2. The scene is composed of a box at the bottom and a cylinder and a small box on the top (see top row in Figure 6.9). We capture the two defocused images shown in the top row in Figure 6.9 and obtain the region Ω_+ shown in the bottom-left image and the depth map in the bottom-right image. The depth map is also displayed as a mesh and viewed from different vantage points in Figure 6.10 to better evaluate the quality of the reconstruction.

6.8 Summary

Defocus can be modeled as a diffusion process, and represented mathematically using the heat equation, where image blur corresponds to diffusion of heat (the radiance, or "deblurred image"). This analogy can be extended to non planar scenes by allowing a space-varying diffusion coefficient.

Although the inverse problem of reconstructing 3-D structure from blurred images corresponds to an "inverse diffusion" that is notoriously ill-posed (see

---

[3]In the experiments, the average does not include the boundary of Ω, $\partial\Omega$, which is three pixels wide.

Figure 6.5. Slope data set. Top-left: Near-focused original image. Top-right: Far-focused original image. Bottom-left: True region Ω_+. Bottom-right: Estimated region Ω_+.

Example F.1 in Appendix F), this problem can be bypassed by using the notion of *relative blur*. Given two images, within each neighborhood, one of the two is sharper than the other, and the amount of diffusion it takes to transform the sharper image into the blurrier one depends on the depth of the scene. This can be used to devise an algorithm that estimates a depth map without recovering the deblurred image, using only forward diffusion.

The reader should tackle this chapter along with Appendix B in order to become familiar with the calculations that are necessary to devise iterative gradient-based algorithms.

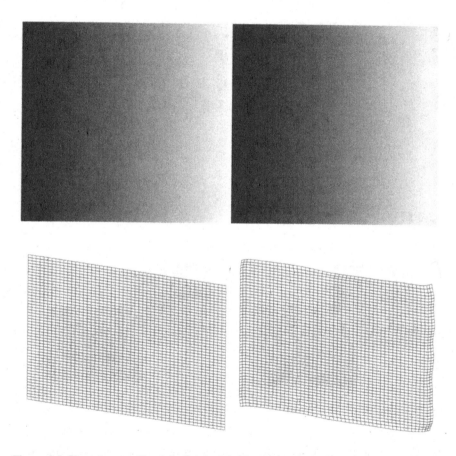

Figure 6.6. Slope data set. Top-left: Original depth map in grayscale intensities. Top-right: Estimated depth map in grayscale intensities. Bottom-left: Original depth map displayed as a mesh. Bottom-right: Estimated depth map displayed as a mesh.

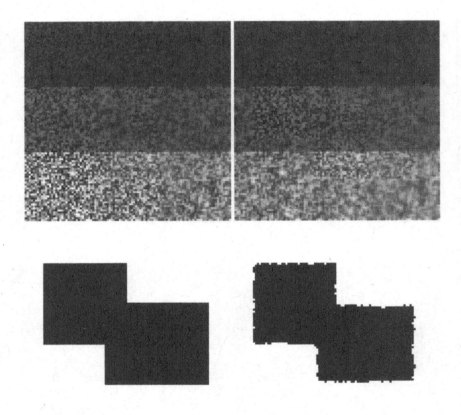

Figure 6.7. Piecewise-constant data set. Top-left: Near-focused original image. Top-right: Far-focused original image. Bottom-left: True region Ω_+. Bottom-right: Estimated region Ω_+.

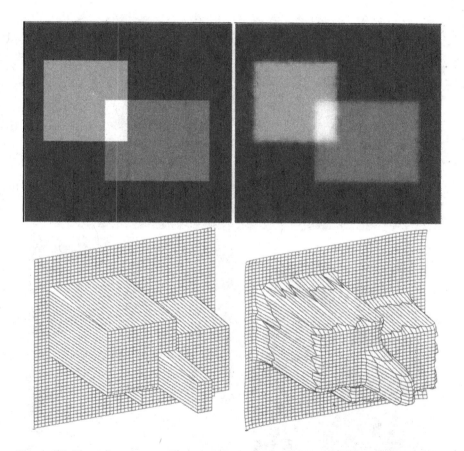

Figure 6.8. Piecewise-constant data set. Top-left: Original depth map in grayscale intensities. Top-right: Estimated depth map in grayscale intensities. Bottom-left: Original depth map displayed as a mesh. Bottom-right: Estimated depth map displayed as a mesh.

Figure 6.9. Real data set. Top-left: Near-focused original image. Top-right: Far-focused original image. Bottom-left: Estimated region Ω_+. Bottom-right: Estimated depth map.

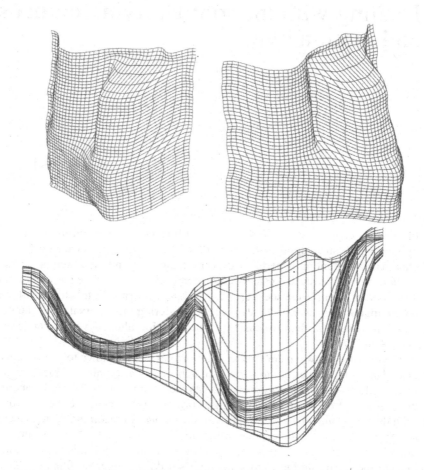

Figure 6.10. Real data set. Top-left: Estimated depth map viewed from the right. Top-right: Estimated depth map viewed from the left. Bottom: Estimated depth map viewed from the top.

7

Dealing with motion: Unifying defocus and motion blur

In previous chapters we have assumed that images are obtained instantaneously by computing an integral of the type (2.4). In practice, physical sensors count photons, and counting is performed over a temporal window, which is called the *shutter interval*, because it usually corresponds to the time while the shutter of the lens remains open. If the scene is static and the camera still, this has no ill effect on the image. However, if the camera moves during this interval, or if the scene moves relative to it, then we can experience motion blur, as we have anticipated in Section 2.5.

The interplay between defocus and motion blur is well known to photographers. One wants a large aperture to allow as much energy as possible to hit the sensor, which reduces the signal-to-noise ratio. On the other hand, the larger the aperture, the smaller the depth of field,[1] so large portions of the scene are blurred out. To limit defocus while maintaining constant the amount of incoming light, one could use a small aperture and keep the shutter mechanism open for a longer time (i.e., use a longer exposure). Unfortunately, if the scene is not still this causes motion-blur artifacts. So, motion (temporal) blur and defocus (spatial blur) are closely related; one can only improve one at the expense of the other. To make such an interplay more explicit and clear, we capture both phenomena with one model in this chapter. We show how to unify the treatment of defocus and motion-blur artifacts, and how to exploit them to reconstruct the shape, radiance, and even the motion of the scene.

---

[1] The depth of field is the distance in front of and behind the object which appears to be in focus.

We consider a scene that is moving rigidly relative to the camera, or a camera that is moving inside a static scene (multiple independently moving objects are addressed in the next chapter). We consider both the problems of reconstructing the 3-D structure and of restoring the images by reducing motion-blur and defocus artifacts. In particular, the analysis of motion blur has become increasingly important in recent years in video processing, where one is interested in obtaining high-resolution still images from video.

However, here we do not analyze single images, because that does not allow us to infer anything about the 3-D structure and motion of the scene in general. Instead, we consider two or more images captured with different focus settings but with the same exposure and, provided that certain assumptions are satisfied, show how they can be used to infer the 3-D structure of the scene, its motion, and its radiance, that can be thought of as a "deblurred" or restored image.

The reader may wish to revisit Section 2.5 as well as 3.5 and Appendix F before delving into the material in this chapter.

7.1   Modeling motion blur and defocus in one go

In Section 2.5.1 we explained the phenomenon of motion blur as the temporal averaging of the photon count while the shutter is open. There we introduced the *velocity field* (2.45) induced by either a moving camera imaging a static scene, or by a rigidly moving scene. If one were given such a field $\nu : \Omega \times [0, \infty) \to \mathbb{R}^2$ then the image would be obtained by integrating the radiance on a temporal window, which can be approximated by a kernel, sort of a "soft temporal window" [2]

$$I(y) = \int \frac{1}{\sqrt{2\pi \Delta \tilde{T}^2}} e^{-t^2/(2\Delta \tilde{T}^2)} r(y + \nu(y, t)) dt, \qquad (7.1)$$

where $\Delta \tilde{T}$ plays the role of the shutter opening interval although, because the shutter opens and closes gradually, one may have to multiply by a constant factor to match the specifications of the camera, as we have discussed in Section 2.5.1.

If we approximate the motion of the camera relative to the scene in (2.41) by a simple fronto-parallel translation $V \in \mathbb{R}^3$ with $V_3 = 0$, as we have done in (2.46), and we further assume that translational velocity is constant in time, we get that the velocity field $\nu^i(y, t)$ induced on the ith image satisfies

$$\nu^i(y, t) = v_i \frac{V_{1,2}}{s(y)} t \qquad \forall \, y \in \Omega, \qquad (7.2)$$

where $V_{1,2}$ is the vector made of the first two components of the velocity V and v_i is the focus setting of the ith blurred image. By direct substitution into equation (7.1), we arrive at the following model for the intensity of a motion-blurred

---

[2]This temporal window accounts for the shutter opening and closing gradually, rather than assuming instantaneous on/off of the shutter mechanism.

image,

$$I(y, v_i) = \int \frac{1}{\sqrt{2\pi}} e^{-t^2/2} r \left(y + \Delta \tilde{T} v_i \frac{V_{1,2}}{s(y)} t \right) dt, \qquad (7.3)$$

which can be rewritten as

$$I(y, v_i) = \int \frac{1}{2\pi |\Sigma_i(y)|} e^{-((y-x)^T \Sigma_i^{-1}(y)(y-x))/2} r(x) dx, \qquad (7.4)$$

where the *variance tensor* $\Sigma_i(y) = (v_i^2 \Delta \tilde{T}^2) / (s^2(y)) V_{1,2} V_{1,2}^T$ and $|\Sigma_i(y)|$ is the determinant of $\Sigma_i(y)$. As we have seen in Section 2.5, blurring due to defocus can also be accounted for by defining the variance tensor $\sigma_i^2(y) I_d$, so that the two effects (defocus and motion blur) can be taken into account simultaneously by choosing a new tensor that is just the sum of the two: $\Sigma_i(y) = \sigma_i^2(y) I_d + (v_i^2 \Delta \tilde{T}^2 / s^2(y)) V_{1,2} V_{1,2}^T$, where I_d denotes the 2×2 identity matrix.

This unified treatment of defocus and motion blur allows us to tap into the analogy, discussed in Section 2.4, of blur as a result of a diffusion. As we did then, we notice that we can represent the image equivalently as the integral in (7.4), or by the differential equation that it solves, which in this case is inhomogeneous and anisotropic:

$$\begin{cases} \dot{u}(x, t) = \nabla \cdot (D_i(x) \nabla u(x, t)) & t \in (0, \infty) \\ u(x, 0) = r(x). \end{cases} \qquad (7.5)$$

As the reader can verify by direct substitution, the image in (7.4) is the solution of the equation

$$I(y, v_i) = u(y, t_1) \qquad (7.6)$$

at a particular time $t_1 = 1/4$ starting from the radiance r as initial condition and using the diffusion tensor

$$D_i(x) = \frac{1}{4t_1} \Sigma_i(x) = \sigma_i^2(x) I_d + \frac{v_i^2 \Delta \tilde{T}^2}{s^2(y)} V_{1,2} V_{1,2}^T. \qquad (7.7)$$

So, equations (7.5) and (7.4) are equivalent (in fact, dual) representations whenever the surface is fronto-parallel. They are only approximately equivalent otherwise.

Note that this model includes standard defocus as a special case, when the second term in the diffusion tensor is zero. Before generalizing the reconstruction algorithms derived in Chapter 6 for the case of defocus to the combination of defocus and motion blur, for due diligence we pause to observe that the differential equation representation of motion blur yields a well-posed model. The reader who trusts our word can move on to the following section without loss of continuity.

7.2   Well-posedness of the diffusion model

Because we consider model (7.5) only within a compact set Ω (the image domain), we introduce boundary conditions in the form of:

$$D_i(x)\nabla u(x,t) \cdot n(x) = 0 \qquad \forall x \in \partial\Omega, \qquad (7.8)$$

where $n(x)$, $x \in \partial\Omega$, denotes the unit vector orthogonal to the boundary of Ω. The more mathematically inclined readers will question whether the partial differential equation (7.5) with such boundary conditions always has a solution. We address this issue in this section, which can be skipped without loss of continity by those readers more interested in the implementation of reconstruction algorithms.

The first step in the analysis is to verify that the parameter-to-output map $(r, s, V_{1,2}) \mapsto u(., t_1)$ is well defined. This follows from results for the degenerate parabolic initial boundary value problems

$$\begin{cases} \dot{u}(x,t) = \nabla \cdot (D_i(x)\nabla u(x,t)) & t \in (0,\infty) \\ u(x,0) = r(x) \\ D_i(x)\nabla u(x,t) \cdot n(x) = 0 & \forall x \in \partial\Omega \end{cases} \qquad (7.9)$$

for diffusion tensors of the form $D_i(x) = \sigma_i(x)^2 I_d + (v_i^2 \Delta \tilde{T}^2)/(s^2(x))V_{1,2}V_{1,2}^T$. As mentioned above, $n(x)$, $x \in \partial\Omega$, denotes the unit vector orthogonal to the boundary of Ω. We assume that the depth map is bounded from above and below; that is,

$$0 < s(x) < \infty, \qquad \forall x \in \Omega. \qquad (7.10)$$

The following result should quell the anxiety of the analytically minded reader, at least for what concerns the existence of weak solutions for the direct problem.

Proposition 7.1. *Let* $r \in L^2(\Omega)$ *and*[3] $s \in \mathbb{H}^1(\Omega)$ *satisfy* (7.10). *Then, there exists a unique weak solution* $u \in C(0, t_1; L^2(\Omega))$ *of* (7.9), *satisfying*

$$\int_0^{t_1} \int_\Omega \lambda_i(x)|\nabla u(x,t)|^2 \, dx \, dt \leq \int_\Omega r(x)^2 \, dx, \qquad (7.11)$$

where $\lambda_i(x) \geq 0$ *denotes the minimal eigenvalue of* $D_i(x)$.

Proof. See Section C.5. $\square$

A second step in the analysis concerns uniqueness. Is a blurred image I generated by a unique pair of radiance r and velocity field ν? We can easily see that this is not the case by considering the simple scenario where the velocity field is constant in time and space and images are affected only by motion blur. In this case, there are infinitely many pairs $\{\hat{r}, \hat{\nu}\}$ that generate the same image I. For example, $\{\hat{r}, \hat{\nu}\} = \{I, 0\}$ and $\{\hat{r}, \hat{\nu}\} = \{r, \nu\}$ are both valid pairs. More in general,

---

[3]The symbol $\mathbb{H}$ denotes the Hausdorff space [Munkres, 1999].

the following is also a valid (infinite) set of pairs:

$$\hat{r} = \int \frac{1}{\sqrt{2\pi}} e^{-t^2/2} r \left(x + \sqrt{1 - \alpha^2} \nu t \right) dt$$

$$\hat{\nu} = \alpha \nu$$

(7.12)

for all $\alpha \in [0, 1]$. This lack of uniqueness reflects the lack of observability of the unknowns (3-D shape and radiance) from the data, an issue that has been addressed in some detail in Chapter 3 for the case of defocus. Unless we have strong priors on the unknowns, it is not possible to meaningfully choose a unique solution among the set of valid ones. Even when we have strong priors, the solution chosen will reflect the choice of priors, which may or may not be meaningful. [4] In order to make an unobservable model observable, either additional data or additional assumptions must be made. This is the approach that we take in the next section.

7.3   Estimating Radiance, Depth, and Motion

The analysis in the previous section, other than placing the analogy of the integral model of the image (7.4) and its differential counterpart (7.5) on more solid ground, serves to convince us that, if we want to recover radiance, depth, and motion, we need to collect more than one defocused, motion-blurred image, and introduce additional constraints.

To this end, we can envision collecting a number K of defocused and motion-blurred images $\{I(\cdot, v_1), \ldots, I(\cdot, v_K)\}$ using in each image a different focus setting $\{v_1, \ldots, v_K\}$. Notice that each of the focus settings $\{v_i\}_{i=1,..,K}$ affects both the isotropic component of the diffusion tensor D_i (defocus) and its anisotropic component $(v_i^2 \Delta \tilde{T}^2 / s^2(x)) V_{1,2} V_{1,2}^T$ (motion blur). As we have discussed in Section 7.1, we can represent an image $I(\cdot, v_i)$ as the solution u_i of equation (7.5) at time $t = t_1 = 1/4$ with diffusion tensor $D_i(x) = \sigma_i^2(x) I_d + (v_i^2 \Delta \tilde{T}^2 / s^2(x)) V_{1,2} V_{1,2}^T$, and with the initial condition $u_i(x, 0) = r(x)$, $\forall i = 1 \ldots K$.

The problem of inferring the radiance r, the depth map s, and the velocity field ν of the scene can be posed by seeking a solution to (7.5) that is as close as possible to the measured images, according to some data fidelity criterion. However, the problem has to be properly regularized in order to make it well-posed. The reader may want to revisit Section 3.5 as well as Appendix F at this point. [5] Those sections should be sufficient to convince the reader that a suitable

---

[4] As an illustrating example, if we try to determine the two factors a and b from their product $c = ab$, we need to know more about the factors. If, say, a, b, and c are square matrices and a is an orthogonal matrix, then b can be recovered uniquely from a Q–R decomposition. However, if a is a general matrix, any assumption imposed to recover a unique b will introduce a spurious structure that has nothing to do with the data, c. For instance, $a = I$ yields $b = c$, but $a = c/2$ yields $b = 2I$.

[5] Additional details on Tikhonov regularization can be found in [Engl et al., 1996].

criterion to minimize must involve a data-fidelity term, say the L^2 norm of the difference between the model image and the measured one, as well as additional terms

$$\hat{r}, \hat{s}, \hat{V}_{1,2} = \arg \min_{r,s,V_{1,2}} \sum_{i=1}^{K} \int_{\Omega} (u_i(x, t_1) - I(x, v_i))^2 \, dx$$
$$+ \alpha_0 \|r - r^*\|^2 + \alpha_1 \|\nabla s\|^2 + \alpha_2 (\|V_{1,2}\| - M_0)^2, \qquad (7.13)$$

weighted by suitable design parameters α_0, α_1, and α_2. Here r^* plays the role of a "prior" for r. Typically we do not have prior information on the radiance r. However, it is necessary to introduce this term to guarantee that the estimated radiance does not grow unbounded. In practice, one can use one of the input images as the prior, or a combination of them, weighted by a very small $\alpha_0 > 0$.

Here M_0 is a positive number related to the maximum amount of motion blur that we are willing to tolerate in the input data. One can choose the norm depending on the desired space of solutions. We choose the L^2 norm for the radiance and the components of the gradient of the depth map and the ℓ^2 norm for $V_{1,2}$.

In this functional, the first term takes into account the discrepancy between the model and the measurements; the second and third terms are classical regularization functionals meant to encourage some regularity in the estimated depth map and penalizing large deviations of the radiance from the prior. The last term may look rather unusual; its main purpose is to exclude the trivial minimum corresponding to $V_{1,2} = 0$. In fact, for $\alpha_2 = 0$ or $M_0 = 0$, $V_{1,2} = 0$ is always a minimum of the functional in (7.13), which is of course undesirable. This minimum also reflects a lack of observability (uniqueness) in the model, as we have discussed at the end of Section 7.2. However, it can be addressed by using generic regularizers, as we do in this chapter, where we transform this minimum into a maximum for positive values of M_0 and α_2 (see Figure 7.1).

7.3.1 Cost Functional Minimization

To minimize the cost functional (7.13) we employ an iterative minimization procedure. We update each unknown incrementally so as to form a sequence that converges to a local minimum of the cost functional. In other words, we design sequences $\hat{r}(x, k)$, $\hat{s}(x, k)$, $\hat{V}_{1,2}(k)$, with $k = 0, 1, \ldots$, such that

$$\hat{r}(x) = \lim_{k \mapsto \infty} \hat{r}(x, k)$$
$$\hat{s}(x) = \lim_{k \mapsto \infty} \hat{s}(x, k) \qquad (7.14)$$
$$\hat{V}_{1,2} = \lim_{k \mapsto \infty} \hat{V}_{1,2}(k).$$

At each iteration we update the unknowns by an increment in the direction opposite to the gradient of the cost functional with respect to the corresponding

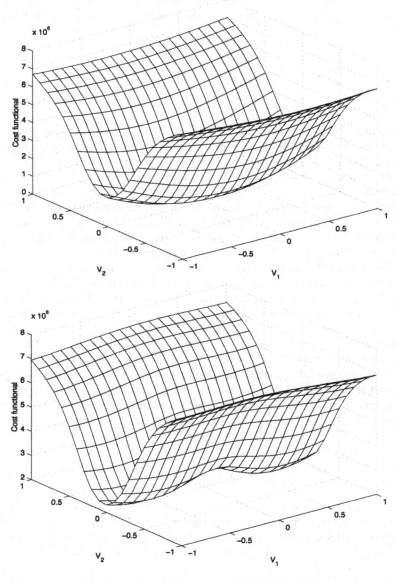

Figure 7.1. Top: Cost functional for various values of V_1 and V_2 when $\alpha_2 = 0$ or $M_0 = 0$. Bottom: Cost functional for various values of V_1 and V_2 when $\alpha_2 \neq 0$ and $M_0 \neq 0$. In both cases the cost functional (7.13) is computed for a radiance $\hat{r}$ and a depth map $\hat{s}$ away from the true radiance r and the true depth map s. Notice that on the bottom plot there are two symmetric minima for $V_{1,2}$. This is always the case unless the true velocity satisfies $V_{1,2} = 0$, because the true $V_{1,2}$ can be determined only up to the sign. Furthermore, notice how the point $V_{1,2} = 0$ is now a maximum.

unknowns:

$$\frac{\partial \hat{r}(x, k)}{\partial k} \doteq -\beta_r \nabla_{\hat{r}} E(x)$$

$$\frac{\partial \hat{s}(x, k)}{\partial k} \doteq -\beta_s \nabla_{\hat{s}} E(x) \qquad (7.15)$$

$$\frac{\partial \hat{V}_{1,2}(k)}{\partial k} \doteq -\beta_V \nabla_{\hat{V}_{1,2}} E.$$

The above iterations decrease the cost functional as k increases as we have shown in equation (5.13). The step sizes β_r, β_s, and β_V can be fixed, as we do in our algorithms, or selected on line by a one-dimensional search (which amounts to the steepest descent method). The computation of the above gradients is explained in detail in Appendix B.2, leading to equation (B.42), which we report below

$$\nabla_r E(x) = 2\sum_{i=1}^{K} w_i(x, t_1) + 2\alpha_0(r(x) - r^*(x))$$

$$\nabla_s E(x) = -2\sum_{i=1}^{K} \int_0^{t_1} \left(\sigma_i(x) \frac{\gamma D v_i}{s^2(x)} I_d - 2\frac{v_i^2 \Delta \tilde{T}^2}{s^3(x)} \left(V_{1,2} V_{1,2}^T \right) \right)$$
$$\nabla u_i(x, t) \cdot \nabla w_i(x, t_1 - t) \, dt - 2\alpha_1 \triangle s(x)$$

$$\nabla_{V_j} E = -2\sum_{i=1}^{K} \int_0^{t_1} \int_\Omega \frac{v_i^2 \Delta \tilde{T}^2}{s^2(x)} \left(e_j V_{1,2}^T + V_{1,2} e_j^T \right)$$
$$\nabla u_i(x, t) \cdot \nabla w_i(x, t_1 - t) \, dx \, dt + 2\alpha_2(|V_{1,2}| - M_0)\frac{V_{1,2}}{|V_{1,2}|},$$
$$(7.16)$$

where $e_1 = [1 \ 0]^T$, $e_2 = [0 \ 1]^T$, and w_i satisfies the following adjoint parabolic equation:

$$\begin{cases} \dot{w}_i(x, t) = \nabla \cdot (D_i(x)\nabla w_i(x, t)) \\ w_i(x, 0) = u_i(x, t_1) - J(x, v_i) \\ (D_i(x)\nabla w_i(x, t)) \cdot n(x) = 0. \end{cases} \qquad (7.17)$$

This provides the basis for the implementation of an algorithm to reconstruct 3-D shape, and restoring a motion-blurred image, which we summarize in Table 7.1. The next section provides some examples of results obtained with such an algorithm.

7.4 Examples

The algorithm sketched in Section 7.3.1 is tested on both synthetic data (Section 7.4.1) and real images (Section 7.4.2). In the first case, we can compare the estimates directly to ground truth and test its performance. In the second case there is no ground truth available, but the reader should still be able to gain an appreciation for the functioning of the algorithm by visual inspection of the results. The gradient flow described in Section 7.3.1 is implemented numerically using standard finite difference schemes.

Table 7.1. Summary of the image restoration and depth map and velocities estimation algorithm from motion-blurred images captured with different focus settings.

Algorithm (motion blur and defocus)

1. Given: Calibration parameters (from knowledge of the camera; see Appendix D for a simple calibration procedure) $v, F, D, \gamma, \Delta\tilde{T}$, K images $\{J(x, v_i)\}_{i=1,...,K}$, a chosen threshold ϵ, regularization parameters α_0, α_1, and α_2, the step size β_r, β_s, and β_V, seek the restored image $\hat{r}$, the depth map $\hat{s}$, and the motion vector $\hat{V}_{1,2}$ as follows.

2. Initialize the radiance $\hat{r}(x, 0)$ to one of the input images; initialize the depth map $\hat{s}(x, 0)$ with a plane at depth

$$\hat{s}(x,0) = \frac{(v_1 + v_2)F}{v_1 + v_2 - 2F};$$

initialize the velocity $\hat{V}_{1,2}(0)$ to a normal vector with random initial direction,

3. Compute the diffusion tensor $\hat{D}_i$ via equation (7.7),

4. Simulate (i.e., numerically evaluate) equation (7.5).

5. Using the solution obtained at the previous step, simulate equation (7.17),

6. Compute the gradient of u and w and evaluate equations (7.16).

7. Update the unknowns via

$$\begin{aligned}
\hat{r}(x, k+1) &= \hat{r}(x, k) - \beta_r \nabla_r E(x) \\
\hat{s}(x, k+1) &= \hat{s}(x, k) - \beta_s \nabla_s E(x) \\
\hat{V}_{1,2}(k+1) &= \hat{V}_{1,2}(k) - \beta_V \nabla_V E.
\end{aligned}$$

8. Return to Step 3 until the norm of the gradients is below the chosen threshold ϵ.

7.4.1 Synthetic Data

In this first set of experiments, we consider a depth map made of a slanted plane (top image in Figure 7.2), that has one side at 520 mm from the camera and the other side at 850 mm from the camera. The slanted plane is painted with a random texture. We define the radiance r to be the image measured on the image plane when a pinhole lens is used (top-left image in Figure 7.3). The top-right image in Figure 7.3 has been captured when the scene or the camera is subject to a sideways translational motion while the camera shutter remains open. Notice that the top portion of the image is subject to a more severe motion blur than the bottom part. This is due to the fact that in this case points that are far from the camera (bottom portion of the image) move at a slower speed than points that are close to the camera (top portion of the image).

Figure 7.2. Top-left: True depth map of the scene. Bottom: Recovered depth map. Top-right: Profile of the recovered depth map. As can be noticed, the recovered depth map is very close to the true depth map with the exception of the top and bottom sides. This is due to the more severe blurring that the input images are subject to at these locations.

We simulate a camera that has focal length 12 mm, F-number 2, and shutter interval $\Delta \tilde{T} = 1$ s. With these settings we capture two images: one by focusing at 520 mm, and the other by focusing at 850 mm. If neither the camera nor the scene is moving, we capture the two images on the bottom row in Figure 7.3. Instead, if either the camera or the scene is moving sideways, we capture the two images on the top row in Figure 7.4. The latter two are the images we give as input to the algorithm. In Figure 7.4 we show the recovered radiance when no motion blur is taken into account (bottom-left image) and when motion blur is taken into account (bottom-right image). As one can notice by inspection, the latter estimate of the radiance is sharper than the estimate of the radiance when motion blur is not modeled. The improvement in the estimate of the radiance can also be evaluated quantitatively because we know the ground truth exactly. To measure the accuracy

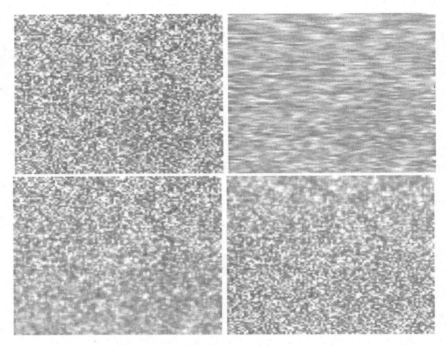

Figure 7.3. Top-left: Synthetically generated radiance. Top-right: Motion-blurred radiance. This image has been obtained by motion-blurring the synthetic radiance on the left. Bottom-row: Defocused images from a scene made from the synthetic radiance in Figure 7.3 (leftmost) and depth map in Figure 7.2 (top-left) without motion blur.

of the estimated radiance, we compute the following discrepancy,

$$Err_1 = \sqrt{E\left[\left(\frac{\hat{\xi}}{\xi} - 1\right)\right]}, \qquad (7.18)$$

where $\hat{\xi}$ is the estimated unknown, ξ is the ground truth, and $E[\cdot]$ denotes the average over Ω. As an example, the discrepancy Err_1 between the true radiance and the motion-blurred radiance (top-right image in Figure 7.3) is 0.2636. When we compensate only for defocus during the reconstruction, the discrepancy Err_1 between the true radiance and the recovered radiance is 0.2642. As expected, because the motion-blurred radiance is the best estimate possible when we do not compensate for motion blur, this estimated radiance cannot be more accurate than the motion-blurred radiance. Instead, when we compensate for both defocus and motion blur, the discrepancy Err_1 between the true radiance and the recovered radiance is 0.2321. This shows that the outlined algorithm can restore images that are not only defocused, but also motion blurred. The recovered depth map is shown in Figure 7.2 on the top-right and bottom images together with the ground

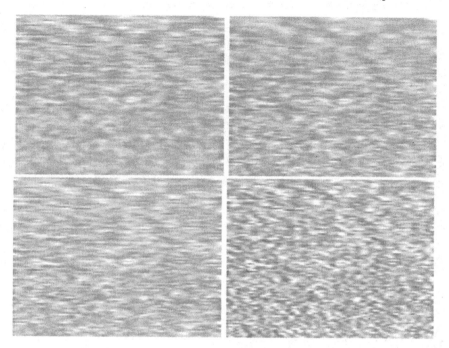

Figure 7.4. Top-row: Defocused (left) and motion-blurred (right) images from a scene made from the synthetic radiance in Figure 7.3 (top-left) and depth map in Figure 7.2 (top-left). Bottom-left: Recovered radiance from the two defocused and motion-blurred images on the left when no motion blur is taken into account ($V_{1,2} = 0$). Bottom-right: Recovered radiance from the two defocused and motion-blurred images on the left when motion blur is taken into account ($V_{1,2} \neq 0$).

truth for direct comparison (top-left). The true motion is $V_{1,2} = [8\ 0]^T$ mm/s and the recovered motion is $[8.079\ -0.713]^T$ mm/s.

7.4.2   Real Images

We test the algorithm on two data sets. The first data set is made of the two real images shown in Figure 7.5. The scene is made of a box that is moving sideways. We capture images with a camera equipped with an AF NIKKOR 35 mm Nikon lens, with F-number 2.8 and shutter interval $\Delta \tilde{T} = 1$ s. We capture the first image by focusing 700 mm from the camera and the second image by focusing 900 mm from the camera. The scene lies entirely between 700 mm and 900 mm. The estimated radiance is shown in Figure 7.5, together with the recovered depth map. The estimated motion is $V_{1,2} = [5.603\ 0.101]^T$ mm/s. In the second data set we use the two defocused and motion-blurred images in Figure 7.6 (first and second images from the left) captured with the same camera settings as in the first data set. The set is composed of a banana and a bagel moving sideways. The

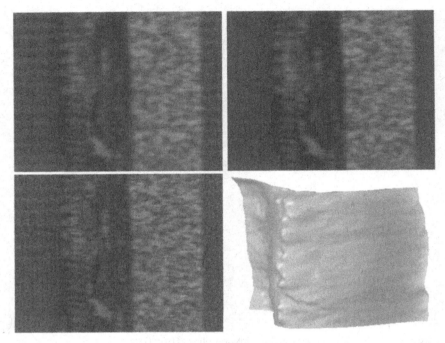

Figure 7.5. Top-row: Input images of the first data set. The two images are both defocused and motion blurred. Motion-blurring is caused by a sideways motion of the camera. Bottom-left: Recovered radiance. Bottom-right: Recovered depth map.

estimated radiance is shown in the third image from the left of the same figure. To visually compare the quality of the estimated radiance, we also add the fourth image from the left in Figure 7.6. This image has been obtained from about the same viewing position when neither the camera nor the scene was moving. Hence, this image is only subject to defocus. The estimated velocity for this data set is $V_{1,2} = [9.639 \ -0.572]^T$ mm/s, that corresponds to a sideways motion.

7.5   Summary

In contrast to defocus, which is a manifestation of a finite (as opposed to infinitesimal) spatial aperture, that is, the area occupied by the lens, motion blur is a manifestation of a finite temporal aperture, that is, the shutter interval. If the scene is rigidly moving relative to the camera during the shutter interval, or if the camera is moving in front of a static scene, the amount of motion blur conveys information about the 3-D shape of the scene as well as its motion.

A motion-blurred image is naturally written as the temporal average during the shutter interval (i.e., as an integral), however, it can also be written, as we did for the case of defocus in the previous chapter, as the solution of a differential

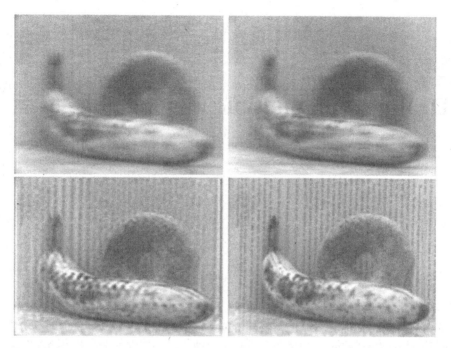

Figure 7.6. Top-row: Input images of the second data set. The two images are both defocused and motion blurred. Motion-blurring is caused by a sideways motion of the camera. Bottom-left: Recovered radiance. Bottom-right: An image taken without motion blur.

equation. This makes it straightforward to unify defocus and motion blur, as we can simply write the (defocused and motion-blurred) image as the solution of a partial differential equation whose diffusion tensor is the sum of two components: an isotropic one due to defocus, and an anisotropic one due to motion blur.

In this chapter we describe an algorithm to infer the depth, radiance, and velocity of a scene from a collection of motion-blurred and defocused images. We have used some simplifying assumptions, for instance, that motion is fronto-parallel translation, although these restrictions are not essential, as one can lift them at the expense of the simplicity of the algorithms.

As we have done in the previous chapter, we have had to add regularization to the cost functionals to make the problem well-posed, following the guidelines developed in Appendix F.

8

Dealing with multiple moving objects

In the previous chapter we have seen how one can exploit motion blur, an artifact of the relative motion between the scene and the camera during the shutter interval, to recover the 3-D structure of the scene along with its motion and the (motion-deblurred) radiance. There we have assumed that there is only one object moving. Either the scene is static and the camera is moving relative to it, or the camera is still and the scene is moving as a single rigid object. In fact, we have further imposed the restriction that motion is just fronto-parallel; that is, the direction of translation is parallel to the image plane.

In this chapter we explore ways in which some of these assumptions can be lifted. The assumption of fronto-parallel translation is not crucial, at least conceptually, because one can always choose more complex scenes that induce more complex image deformations. So, we work to relax the most restrictive assumption, that is, the presence of a single object. We instead assume that there are a number of objects moving with different motions, although we will assume that such motions are, locally in space and time, fronto-parallel translations.

Note that "object" here may not correspond to the intuitive "semantic" notion of objects in the scene, and what an object is may depend on the motion model chosen. For us, an object is a portion of the scene that moves with a coherent motion, according to the chosen model. If we choose a rigid motion model, but the scene contains, say, a human body, then each limb can be considered approximately rigid and constitutes a different object. If we choose a fronto-parallel motion model, and the scene contains, say, a rotating fan, then the scene is broken down into a number of regions each of which can be well approximated by a fronto-parallel motion.

This phenomenon reflects a fundamental modeling tradeoff. We can represent the scene as one single object, and then choose a very complex motion model (including nonrigid deformations) that captures the phenomenology of the scene within a prescribed accuracy. Or, we can choose a simpler class of models (say fronto-parallel translation) and then partition the scene into a number of regions each of which is described by the motion model within the prescribed accuracy. As an analogy, think of approximating a function in an interval: one can choose a prescribed tolerance ϵ, and then choose a parametric class of functions (say polynomials) with order high enough so that the function will be approximated in the entire interval within ϵ. Or, one can choose a simple class of functions (say constant), and break down the interval in such a way that the sum of the errors in all the intervals is less than ϵ. The larger the class of motions, the fewer are the intervals.

In this chapter we also consider playing with the exposure of our camera in order to generate sequences of images that allow us to infer spatial properties about the scene. In the process, we show how to separate the effects of defocus from those of motion blur.

We follow the same line of thought presented in Chapter 6, where we have derived an equivalent image formation model in terms of diffusion. In particular, we introduce the notion of relative motion blur to eliminate the unknown radiance from the inference process. We do so in both the cases when velocities remain constant and when they change direction between different exposures.

Finally, we show how to perform image restoration in the most general case where there are multiple moving objects that could change direction from one image to the next.

8.1   Handling multiple moving objects

In this section we consider scenes that are made of Lambertian (see Appendix A), opaque objects with unknown shape and pose, moving in different directions. For us, an object is determined by a region of the image, and we do not model occlusions explicitly. Indeed, as we show in the next chapter, occlusions in the case of finite-aperture cameras do not simply cause discontinuities in the image intensities, because both objects may be visible in the same region. We address this effect explicitly in the next chapter. For now, we assume that occlusions simply cause discontinuities in the depth map, so we can still represent the 3-D structure of the scene with a function $s : \Omega \mapsto (0, \infty)$ that assigns a depth value to each pixel coordinate. Therefore, each pixel on the image plane is assigned to only one object.

Now, suppose that we have Y objects moving in the scene; then, we can uniquely assign a set of pixels $\Omega_j \subset \Omega$ to the jth object, so that $\{\Omega_j\}_{j=1,\dots,Y}$ is a partition of Ω (see Section 2.5 for more details). Recalling our definition of the velocity field ν in (2.45) we obtain that $\nu(y,t)$ restricted to $y \in \Omega_j$ defines the

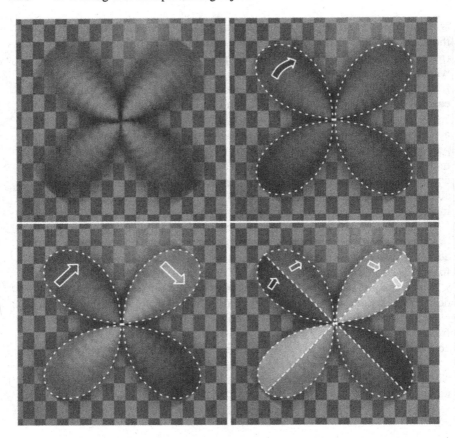

Figure 8.1. Multiple moving objects. This figure illustrates the notion of "object" as a portion of the image undergoing coherent motion. What an object is thus depends on the class of motions chosen (e.g., rigid vs. fronto-parallel) as well as the tolerance, so that under the same model, many more objects are found if the tolerance is small. For instance, consider the propeller in the top-left image. It is a single object that undergoes a rigid motion (top-right). However, by using a fronto-parallel model, the same object might be segmented into 2 objects (bottom-left) or more (bottom-right) depending on the tolerance of the fronto-parallel model.

motion of the jth object. Figure 8.1 illustrates the model for motion and structure that we just introduced, as well as the fact that what an object is depends on the class of motions chosen as well as on the tolerance threshold. As we did in the previous chapter, we consider a 2-D, fronto-parallel translational motion, so that

$$\nu(y,t) = v\frac{V_{1,2}^j}{s(y)}t \quad \forall\, y \in \Omega_j, \qquad (8.1)$$

where v is the focus setting and $V^j_{1,2} \in \mathbb{R}^2$ are the first two components of the 3-D translation of the jth object. Because there is a scale ambiguity between the magnitude of the velocity $V^j_{1,2}$ of a region Ω_j and its depth, visible from equation (8.1) (just multiply V and s by any constant, the resulting velocity field is unchanged) objects that are semantically different but that are moving along the same direction will be indistinguishable and therefore assigned to the same region (Figure 8.1). As we have observed, one could certainly consider motion models that are more sophisticated than equation (8.1). However, a different motion model will simply result in different partitions where the model is satisfied within the prescribed accuracy. In this chapter we are interested in keeping the analysis as simple as possible and, therefore, we choose the simplest fronto-parallel translational model and assume constant velocity during the shutter interval.

As we suggested in Section 2.5, we represent the regions implicitly using signed distance functions [Sethian, 1996]. For simplicity, we consider the case of two regions, Ω_1 and Ω_2, so that a single level set function is sufficient to encode them. As mentioned in Section 2.3, notice how one can make a single-connected component in Figure 2.5, a set composed of two connected components by shifting the level set function ϕ down. The extension to more than two regions can be achieved by considering more level set functions, as shown in [Vese and Chan, 2002]. However, we do not explore this extension here. Recall that the level set function ϕ is a map $\phi : \Omega \mapsto \mathbb{R}$ that has the desired region as its zero level set:

$$\begin{aligned} \Omega_1 &= \{x \in \Omega : \phi(x) \geq 0\} \\ \Omega_2 &= \{x \in \Omega : \phi(x) < 0\} = \Omega \backslash \Omega_1. \end{aligned} \qquad (8.2)$$

Using the Heaviside function H

$$H(z) = \left\{ \begin{array}{ll} 1, & \text{if } z \geq 0 \\ 0, & \text{if } z < 0 \end{array} \right. \qquad (8.3)$$

we can equivalently write

$$\begin{aligned} \Omega_1 &= \{x \in \Omega : H(\phi(x)) = 1\} \\ \Omega_2 &= \{x \in \Omega : H(\phi(x)) = 0\}. \end{aligned} \qquad (8.4)$$

In summary, the jth object is:

- Determined by the region Ω_j of the image (more technically, it is its pre-image under the image formation model)

$$\Omega_j = \left\{ x \in \Omega : H(\phi(x)) = \left\{ \begin{array}{ll} 1 & \text{if } j = 1 \\ 0 & \text{if } j = 2 \end{array} \right. \right\}. \qquad (8.5)$$

- Its geometry is represented by the depth map $s(y)$, $\forall y \in \Omega_j$.

- Its de-blurred image (radiance) is represented by $r(y)$, $\forall y \in \Omega_j$.

- Its motion is 2-D fronto-parallel and with constant velocity during the shutter interval; that is,

$$\nu(y, t) = v \frac{V^j_{1,2}}{s(y)} t \qquad \forall\, y \in \Omega_j. \qquad (8.6)$$

Finally, its image is given by

$$I(y) = \int \frac{1}{\sqrt{2\pi}} e^{-t^2/2} r\left(y + v\Delta T \frac{V_{1,2}^j}{s(y)} t\right) dt \qquad \forall y \in \Omega_j \qquad (8.7)$$

or by its differential counterpart (7.9), where y is restricted to Ω_j. It is this latter form that proves more suitable to extending the algorithms developed in the previous chapter to multiple moving objects.

8.2   A closer look at camera exposure

So far we have exploited changes in focus setting to recover the 3-D shape, radiance, and a simple model of the motion of the scene. We are going to start from the diffusion model of image formation (7.9), where the most crucial component, the diffusion tensor, is repeated here for completeness:

$$D(y) = \sigma^2(y)I_d + \frac{v^2\Delta\tilde{T}^2}{s^2(y)}V_{1,2}V_{1,2}^T, \qquad (8.8)$$

where

$$\sigma^2(y) = \frac{\gamma^2 D^2 v^2}{4}\left(\frac{1}{F} - \frac{1}{v} - \frac{1}{s(y)}\right)^2 \qquad (8.9)$$

is the term associated with defocus, and

$$\frac{v^2\Delta\tilde{T}^2}{s^2(y)}V_{1,2}V_{1,2}^T \qquad (8.10)$$

is the term associated with motion blur. To simplify the analysis, we base most of our arguments solely on the diffusion tensor. Optimality is guaranteed when the scene is an equifocal plane, and yields an approximation otherwise.

If we change the focus setting v, both defocus and motion blur are affected, thus making it difficult to analyze their effects in isolation. However, if we capture images of the same scene from the same vantage point while changing the exposure interval $\Delta\tilde{T}$, then we may be able to separate defocus from motion blur. Before showing how to do so, notice that we assume that images are normalized with respect to the amount of exposure and that there are no saturations or nonlinearities in the camera sensor.

Suppose that we simulate two images $I(y, \Delta\tilde{T}_1)$ and $I(y, \Delta\tilde{T}_2)$ by setting the exposure of the first image to $\Delta\tilde{T}_1$ and of the second to $\Delta\tilde{T}_2$. We are using the notation $I(y, \Delta T)$ to indicate the dependency of the image on the shutter interval ΔT in the same way we did in Chapter 6 to indicate dependency on the focus setting. We assume, without loss of generality, that

$$\Delta\tilde{T}_2 > \Delta\tilde{T}_1. \qquad (8.11)$$

Following the same argument developed in Section 6.2, the reader can see that the difference between the diffusion tensor of the first image and that of the second image is

$$\Delta D(y) \doteq D_2(y) - D_1(y) = \left(\Delta \tilde{T}_2^2 - \Delta \tilde{T}_1^2\right) \frac{v^2 V_{1,2} V_{1,2}^T}{s^2(y)}. \qquad (8.12)$$

As we can see, this new modality lets us separate the two diffusion terms in equation (8.8) by simply evaluating the relative diffusion of images with different exposure. The use of equation (8.12) is the main theme of the next section.

8.3 Relative motion blur

In this section we extend the notion of relative blur introduced in Section 6.2 to motion blur generated by multiple objects. Because we have different objects j and different images i, as well as different intervals ΔT and different focus settings v, the notation tends to become a bit cumbersome. The reader will forgive us if we have to use multiple indices, but unfortunately the complexity of the image formation process forces us to do so.

The reader who has followed the introduction of relative diffusion in Chapter 6 will notice that equation (8.12) begs to be treated in the same way in order to eliminate the unknown radiance. Following the same steps we arrive at an inhomogeneous and anisotropic diffusion model for relative motion blur of the jth object:

$$\begin{cases} \dot{u}_j(x,t) & = \nabla \cdot (\Delta D_j(x) \nabla u_j(x,t)) \\ u_j(x,0) & = I(x, \Delta \tilde{T}_1) \end{cases} \quad t \in (0, \infty), \quad \forall x \in \Omega_j, \quad (8.13)$$

where the second image $I(x, \Delta \tilde{T}_2)$ is the solution of the equation at time t_1

$$I(x, \Delta \tilde{T}_2) = u_j(x, t_1) \qquad (8.14)$$

and the diffusion tensor ΔD_j is defined as

$$\Delta D_j(y) \doteq \left(\Delta \tilde{T}_2^2 - \Delta \tilde{T}_1^2\right) \frac{v^2 V_{1,2}^j \left(V_{1,2}^j\right)^T}{s^2(y)}. \qquad (8.15)$$

Now, given two images $J(x, \Delta \tilde{T}_1)$ and $J(x, \Delta \tilde{T}_2)$ as input, we are ready to formulate the problem of inferring the depth map s, the velocities $\{V_{1,2}^1, V_{1,2}^2\}$, and the partition $\{\Omega_1, \Omega \backslash \Omega_1\}$ of the scene by minimizing the following least-squares functional with Tikhonov regularization (the reader may want to review the basics of regularization in Appendix F)

$$E = \int_{\Omega_1} \left(u_1(x, t_1) - J(x, \Delta \tilde{T}_2) \right)^2 dx$$
$$+ \int_{\Omega \backslash \Omega_1} \left(u_2(x, t_1) - J(x, \Delta \tilde{T}_2) \right)^2 dx$$
$$+ \alpha_0 \|\nabla s\|^2 + \alpha_1 \left(\int_{\Omega} s(x) dx - s^{mean} \right)^2 + \alpha_2 \|\nabla H(\phi)\|^2 ;$$

(8.16)

that is, we seek

$$\hat{\phi}, \hat{s}, \hat{V}_{1,2}^1, \hat{V}_{1,2}^2 = \arg \min_{\phi, s, V_{1,2}^1, V_{1,2}^2} E,$$

(8.17)

where α_0, α_1, and α_2 are positive regularization parameters and s^{mean} is a suitable positive number. [1] One can choose the norm $\|\cdot\|$ depending on the desired space of solutions. We choose the L^2 norm for the radiance and the components of the gradient of the depth map.

In this functional, the first two terms take into account the discrepancy between the model and the measurements; the third term imposes some regularity (smoothness) on the estimated depth map. The fourth term fixes the scale ambiguity between the depth map s and the translational velocities $V_{1,2}^j$. To fix the scale ambiguity we choose the mean of the depth map s to be equal to a constant s^{mean}, so that small changes of s will not result in sensible variations of this term. Finally, the last term imposes a length constraint on the boundary of Ω_1 thus penalizing boundaries that are too fragmented or irregular. To minimize the cost functional (8.17) we employ a gradient descent flow.

8.3.1 Minimization algorithm

As we have anticipated, we assume that there are $Y = 2$ objects, so that we can use a single level set function ϕ to represent both Ω_1 and $\Omega_2 = \Omega \backslash \Omega_1$, and that velocities do not change between the input images (i.e., velocities do not depend on the image index i). Let $\hat{\xi}(x, k)$ represent the iterative sequence (in the parameter k) that converges to one of the cost functional unknowns $\hat{\xi} \in \{\hat{\phi}(x), \hat{s}(x), \hat{V}_{1,2}^1, \hat{V}_{1,2}^2\}$. For each of these unknowns, we compute a sequence converging to a local minimum of the cost functional; that is,

$$\hat{\xi}(x) = \lim_{k \to \infty} \hat{\xi}(x, k).$$

(8.18)

At each iteration we update the unknowns by moving in the opposite direction of the gradient of the cost functional with respect to the unknowns. In other words,

---

[1]As mentioned in Section 2.5.1, there is a scale ambiguity between the velocity field v and the depth map of the scene. We choose to fix this by normalizing the average depth to be a constant s^{mean}.

we let

$$\frac{\partial \hat{\xi}(x,k)}{\partial k} \doteq -\nabla_{\hat{\xi}} E(x), \qquad (8.19)$$

where $\nabla_{\hat{\xi}} E(x)$ is the gradient of the cost functional with respect to $\hat{\xi}$. It was shown in equation (5.13) that the above iterations decrease the cost functional as k increases. The computation of the above gradients is involved due to the fact that the explicit solution u of equation (8.13) is not available, and we detail the calculations in Appendix B. Here we only report the final formula from equation (B.57)

$$\begin{aligned}
\nabla_s E(x) &= -2H(\phi(x)) \int_0^{t_1} (B_1(x)\nabla u_1(x,t)) \cdot \nabla w_1(x, t_1 - t)\, dt \\
&\quad - 2H(-\phi(x)) \int_0^{t_1} (B_2(x)\nabla u_2(x,t)) \cdot \nabla w_2(x, t_1 - t)\, dt \\
&\quad + 2\alpha_1 \left(\int_\Omega s(y)dy - s^{mean} \right) - 2\alpha_0 \Delta s(x) \\
\nabla_{V_{1,2}^1} E &= -2 \int_0^{t_1} \int_{\Omega_1} (C_1(x)\nabla u_1(x,t)) \cdot \nabla w_1(x, t_1 - t)\, dx\, dt \\
\nabla_{V_{1,2}^2} E &= -2 \int_0^{t_1} \int_{\Omega \backslash \Omega_1} (C_2(x)\nabla u_2(x,t)) \cdot \nabla w_2(x, t_1 - t)\, dx\, dt \\
\nabla_\phi E(x) &= \delta(\phi(x)) (F_1(x) - F_2(x)) - 2\alpha_2 \delta(\phi(x)) \nabla \cdot \left(\frac{\nabla \phi(x)}{\|\nabla \phi(x)\|} \right),
\end{aligned}$$
$$(8.20)$$

where for $j = 1, 2$,

$$\begin{aligned}
B_j(x) &= -2 \frac{v^2 \left(\Delta \tilde{T}_2^2 - \Delta \tilde{T}_1^2 \right)}{s^3(x)} V_{1,2}^j (V_{1,2}^j)^T \\
C_j(x) &= 2 \frac{v^2 \left(\Delta \tilde{T}_2^2 - \Delta \tilde{T}_1^2 \right)}{s^2(x)} \left(\begin{bmatrix} 1 \\ 1 \end{bmatrix} (V_{1,2}^j)^T + V_{1,2}^j [1\ 1] \right) \\
F_j(x) &= (u_j(x, t_1) - J(x, \Delta \tilde{T}_2))^2,
\end{aligned}$$
$$(8.21)$$

and $w_j(x,t)$ satisfies the following adjoint parabolic equation

$$\left\{ \begin{array}{l}
\dot{w}_j(x,t) = \nabla \cdot (D_j(x)\nabla w_j(x,t)) \\
w_j(x,0) = u_j(x,t_1) - J(x, \Delta \tilde{T}_2) \\
(D_j(x)\nabla w_j(x,t)) \cdot n(x)_j = 0 \qquad \forall x \in \partial \Omega_j
\end{array} \right. \qquad (8.22)$$

with diffusion tensor D_j from equation (8.15).

We summarize the algorithm in Table 8.1.

8.4 Dealing with changes in motion

In this section we analyze in more detail the case of motion blur restoration from multiple images when motion of objects may change both magnitude and direc-

Table 8.1. Summary of the depth map and velocities estimation and object localization algorithm via relative diffusion.

Algorithm (relative motion blur)

1. Given: Calibration parameters (from knowledge of the camera; see Appendix D for a simple calibration procedure) $v, F, D, \gamma, \Delta\tilde{T}_1, \Delta\tilde{T}_2$, two images $J(x, \Delta\tilde{T}_1)$, $J(x, \Delta\tilde{T}_2)$, a chosen threshold ϵ, regularization parameters $\alpha_0, \alpha_1, \alpha_2$, the scale parameter s^{mean}, and step size $\beta_s, \beta_{V^1}, \beta_{V^2}$, and β_ϕ seek the depth map $\hat{s}$, motion vectors $\hat{V}^1_{1,2}, \hat{V}^2_{1,2}$, and level set $\hat{\phi}$ as follows.

2. Initialize depth map $\hat{s}(x, 0)$ with a plane at depth

$$\hat{s}(x, 0) = \frac{(v_1 + v_2)F}{v_1 + v_2 - 2F};$$

initialize level set $\hat{\phi}(x, 0)$ so that it is the signed distance function of uniformly distributed circles on the image domain; initialize velocities $\hat{V}^1_{1,2}(0)$, $\hat{V}^2_{1,2}(0)$ to two orthonormal vectors with random initial direction.

3. Compute the relative diffusion tensors ΔD_j via equation (8.15),

4. Simulate (i.e. numerically evaluate) equation (8.13),

5. Using the solution obtained at the previous step, simulate equation (8.22),

6. Compute the gradient of u and w and evaluate equations (8.20),

7. Update the unknowns via

$$
\begin{aligned}
\hat{s}(x, k+1) &= \hat{s}(x, k) - \beta_s \nabla_s E(x) \\
\hat{V}^1_{1,2}(k+1) &= \hat{V}^1_{1,2}(k) - \beta_{V^1} \nabla_{V^1} E \\
\hat{V}^2_{1,2}(k+1) &= \hat{V}^2_{1,2}(k) - \beta_{V^2} \nabla_{V^2} E \\
\hat{\phi}(x, k+1) &= \hat{\phi}(x, k) - \beta_\phi \nabla_\phi E(x).
\end{aligned}
$$

8. Return to Step 3 until norm of gradients is below the chosen threshold ϵ.

tion from one image to another. As has been done before, in Chapter 6 and in Section 8.3, we are interested in finding a way to avoid image restoration while inferring 3-D shape and motion from motion-blurred images. We show that it is possible to design such an approach, by simultaneously diffusing pairs of images at a time. Also, we analyze the new inference problem and show that it has the same set of solutions (i.e., ambiguities) as the original problem, where one has to recover the radiance as well.

8.4.1   Matching motion blur along different directions

To keep the presentation simple, let us assume that $Y = 1$ (i.e., that a single object is moving in the scene), and that $K = 2$ (i.e., that we have only two images as input, obtained by changing the lone exposure interval). Recall that the image formation model for a motion-blurred image $I(x, \Delta \tilde{T}_i)$, $i = 1, \ldots, 2$, is

$$\begin{cases} \dot{u}_i(x,t) = \nabla \cdot (D_i(x) \nabla u_i(x,t)) & \forall t \in (0, \infty), \forall x \in \Omega \\ u_i(x,0) = r(x) & \forall x \in \Omega \\ (D_i(x) \nabla u_i(x,t)) \cdot n(x) = 0 & \forall x \in \partial\Omega \end{cases} \tag{8.23}$$

with

$$D_i(x) = \sigma^2(x) I_d + v^2 \Delta \tilde{T}_i^2 \frac{V_{1,2}^i \left(V_{1,2}^i\right)^T}{s^2(x)} \qquad \forall x \in \Omega, \tag{8.24}$$

and $n(x)$ with $x \in \partial\Omega$ denotes the unit vector normal to the boundary of Ω. Notice that only the exposure interval $\Delta \tilde{T}_i$ and the velocity vectors $V_{1,2}^i$ in the diffusion tensor may vary from one image to another.

Now, if we change the initial conditions of equation (8.23) to the first input image $I(y, \Delta \tilde{T}_1)$, and then seek to match the solution of equation (8.23) at time t_1 to the other input image, the resulting relative diffusion is

$$\Delta D(x) = \frac{v^2}{s^2(x)} \left(\Delta \tilde{T}_2^2 V_{1,2}^2 \left(V_{1,2}^2\right)^T - \Delta \tilde{T}_1^2 V_{1,2}^1 \left(V_{1,2}^1\right)^T\right). \tag{8.25}$$

In general, $\Delta D(x)$ is neither a positive-(semi)definite nor a negative-(semi)definite tensor, due to the fact that the velocities $V_{1,2}^1$ and $V_{1,2}^2$ are not parallel. Hence, we cannot use this approach as it would result in an unstable model.

Rather, if we consider two diffusion models, one with the first input image $J(y, \Delta \tilde{T}_1)$ as initial condition,

$$\begin{cases} \dot{u}_1(x,t) = \nabla \cdot \left(\hat{D}_1(x) \nabla u_1(x,t)\right) & \forall t \in (0, \infty), \forall x \in \Omega \\ u_1(x,0) = J(x, \Delta \tilde{T}_1) & \forall x \in \Omega \\ (\hat{D}_1(x) \nabla u_1(x,t)) \cdot n(x) = 0 & \forall x \in \partial\Omega \end{cases} \tag{8.26}$$

and the other with the other input image as initial condition,

$$\begin{cases} \dot{u}_2(x,t) = \nabla \cdot \left(\hat{D}_2(x) \nabla u_2(x,t)\right) & \forall t \in (0, \infty), \forall x \in \Omega \\ u_2(x,0) = J(x, \Delta \tilde{T}_2) & \forall x \in \Omega \\ (\hat{D}_2(x) \nabla u_2(x,t)) \cdot n(x) = 0 & \forall x \in \partial\Omega \end{cases} \tag{8.27}$$

with

$$\hat{D}_i(x) = \hat{W}^i(x) \left(\hat{W}^i(x)\right)^T \qquad i = 1, 2 \tag{8.28}$$

where $\hat{W}^i : \mathbb{R}^2 \to \mathbb{R}^2$ is a 2-D vector field, then we may seek to match the solution u_1 and u_2 at time t_1 of these two diffusion models. This condition amounts to having $\forall x \in \Omega$,

$$B_1 + \hat{W}^2(x)\left(\hat{W}^2(x)\right)^T = B_2 + \hat{W}^1(x)\left(\hat{W}^1(x)\right)^T, \qquad (8.29)$$

where for $i = 1, 2$,

$$B_i = v^2 \Delta \tilde{T}_i^2 \frac{V_{1,2}^i \left(V_{1,2}^i\right)^T}{s^2(x)}. \qquad (8.30)$$

By direct substitution one can see that the following are admissible solutions

$$\hat{W}^1(x) = \sqrt{b^2 + 1}v\Delta\tilde{T}_1 \frac{V_{1,2}^1}{s(x)} + bv\Delta\tilde{T}_2 \frac{V_{1,2}^2}{s(x)} \qquad (8.31)$$

$$\hat{W}^2(x) = bv\Delta\tilde{T}_1 \frac{V_{1,2}^1}{s(x)} + \sqrt{b^2 + 1}v\Delta\tilde{T}_2 \frac{V_{1,2}^2}{s(x)}, \qquad (8.32)$$

where $b \in \mathbb{R}$. To make sure that these are all the admissible solutions, we start in all generality by letting

$$\hat{W}^1(x) = av\Delta\tilde{T}_1 \frac{V_{1,2}^1}{s(x)} + bv\Delta\tilde{T}_2 \frac{V_{1,2}^2}{s(x)} \qquad (8.33)$$

$$\hat{W}^2(x) = cv\Delta\tilde{T}_1 \frac{V_{1,2}^1}{s(x)} + dv\Delta\tilde{T}_2 \frac{V_{1,2}^2}{s(x)}, \qquad (8.34)$$

where a, b, c, and d are scalars. Because $V_{1,2}^1$ and $V_{1,2}^2$ are not parallel by assumption (i.e., motion direction changes from one image to another), the above definitions are the most general ones for the unknown velocity fields $\hat{W}^1$ and $\hat{W}^2$. By substituting the above equations in equation (8.29) we obtain that

$$\begin{aligned} ab &= cd \\ a^2 &= c^2 + 1 \\ d^2 &= b^2 + 1, \end{aligned} \qquad (8.35)$$

which can be simplified as

$$\begin{aligned} d &= \pm a \\ c &= \pm b \\ a^2 &= b^2 + 1. \end{aligned} \qquad (8.36)$$

We now notice that a change in the sign of the velocity field does not affect motion blur due to the symmetry of the blurring kernel. In our computations this means that we can choose $a = \sqrt{b^2 + 1}$ to be positive, and this results in both $c = b$ and $d = \sqrt{b^2 + 1}$, which yields the solutions in equations (8.31) and (8.32). This means that in order to have a unique solution during the reconstruction process, we need a strategy to choose one of the solutions. Because we only have access to the velocity fields $\hat{W}^1$ and $\hat{W}^2$, we could choose to set b as the value that minimizes

$$\|\hat{W}^1\|^2 + \|\hat{W}^2\|^2. \qquad (8.37)$$

The minimum is unique (up to a sign [2]) and can be computed in closed form as

$$b = \sqrt{-\frac{1}{2} + \frac{1}{2\sqrt{1 - 4b_0^2}}}, \tag{8.38}$$

where

$$b_0 = \frac{\Delta\tilde{T}_1 \Delta\tilde{T}_2 \left(V_{1,2}^1\right)^T V_{1,2}^2}{\Delta\tilde{T}_1^2 \|V_{1,2}^1\|^2 + \Delta\tilde{T}_2^2 \|V_{1,2}^2\|^2}. \tag{8.39}$$

However, rather than employing the expression above of b in (8.38), we choose to impose the same constraint by adding equation (8.37) as a soft term to the cost functional. By doing so we limit the nonlinearities in the gradients of the cost functional with respect to $V_{1,2}^1$ and $V_{1,2}^2$. Hence, in our estimation problem we select a unique solution by also minimizing equation (8.37). Given $Y = 1$ (i.e., there is a single moving object) and $K = 2$ images $J(\cdot, \Delta\tilde{T}_1)$ and $J(\cdot, \Delta\tilde{T}_2)$ are collected while the shutter remains open for different spans of time $\Delta\tilde{T}_1 < \Delta\tilde{T}_2$, one can pose the problem of inferring the depth s, velocities $\hat{V}_{1,2}^1$ and $\hat{V}_{1,2}^2$, and the parameter $\hat{b}$ via the following minimization,

$$\hat{s}, \hat{V}_{1,2}^1, \hat{V}_{1,2}^2, \hat{b} = \arg \min_{s, V_{1,2}^1, V_{1,2}^2, b} E, \tag{8.40}$$

where

$$\begin{aligned} E \;=\; & \int_\Omega (u_1(x, t_1) - u_2(x, t_1))^2 dx \\ & + \alpha_0 \|\nabla s\|^2 \\ & + \alpha_1 \left(\int_\Omega s(x)dx - s^{mean}\right)^2 \\ & + \alpha_2 \left(\|\hat{W}^1\|^2 + \|\hat{W}^2\|^2\right). \end{aligned} \tag{8.41}$$

8.4.2   A look back at the original problem

At this point, one may wonder whether the new formulation in equation (8.41), for $\alpha_0 = \alpha_1 = \alpha_2 = 0$ and noise-free images, introduces additional ambiguities in the original problem

$$\hat{V}_{1,2}^1, \hat{V}_{1,2}^2, \hat{r}, \hat{s} \;=\; \arg \min_{V_{1,2}^1, V_{1,2}^2, r, s} \sum_{i=1}^{2} \int_\Omega (u_i(x, t_1) - I(x, \Delta\tilde{T}_i))^2 dx, \tag{8.42}$$

---

[2]Note that choosing b with opposite sign as in equation (8.38) implies a change in sign of either $V_{1,2}^1$ or $V_{1,2}^2$. Similarly, both $V_{1,2}^1$ and $V_{1,2}^2$ can change sign simultaneously, without affecting motion blur. All these solutions are isolated, so they do not affect the stability of our gradient-based minimization method.

where u_i is given by equation (8.23) and one also restores the radiance r. To verify that this is not the case we explicitly derive the solution to the original problem by solving

$$u_i(x, t_1) = I(x, \Delta \tilde{T}_i) \qquad i = 1, 2 \qquad (8.43)$$

because at least one such solution exists (notice that now we consider noise-free images). If we define the unknown radiance $\hat{r}$ as

$$\hat{r}(x) = w(x, t_1), \qquad (8.44)$$

where

$$\begin{cases} \dot{w}(x, t) = \nabla \cdot (G(x) \nabla w(x, t)) & \forall t \in (0, \infty), \forall x \in \Omega \\ w(x, 0) = r(x) & \forall x \in \Omega \\ (G(x) \nabla w(x, t)) \cdot n(x) = 0 & \forall x \in \partial\Omega \end{cases} \qquad (8.45)$$

and

$$\begin{aligned} G(x) \;=\; & -\frac{v^2 \Delta \tilde{T}_1^2}{s^2(x)} \left(b^2 V_{1,2}^1 \left(V_{1,2}^1 \right)^T + b^2 V_{1,2}^2 \left(V_{1,2}^2 \right)^T \right. \\ & \left. + b\sqrt{1 + b^2} \left(V_{1,2}^2 \left(V_{1,2}^1 \right)^T + V_{1,2}^1 \left(V_{1,2}^2 \right)^T \right) \right) \quad \forall x \in \Omega \end{aligned} \qquad (8.46)$$

and choose $\hat{W}^1$ and $\hat{W}^2$ to be given from equations (8.31) and (8.32), then the solutions of equation (8.23) with initial conditions $\hat{r}$ and diffusion tensors as in equation (8.24) also satisfy[3] equation (8.43). Because every solution of equation (8.43) satisfies the image models (8.26) and (8.27) by construction, then the original problem and the one formulated in the previous section are equivalent.

Remark 8.1. *A direct consequence of this analysis is that an image restoration (and shape estimation) algorithm where one uses as input a pair of images that are motion blurred along different directions cannot retrieve the original motion vectors. To make the algorithm numerically well behaved, one has to introduce some additional constraints to the estimation problem. For example, one can add the term in equation (8.37) to the functional in equation (8.42) as we did in the previous section.*

8.4.3 Minimization algorithm

Again we follow the trace of the previous sections to minimize the cost functional above. Let $Y = 1$, and $\{\hat{s}(x), \hat{V}_{1,2}^1, \hat{V}_{1,2}^2, \hat{b}\}$ be the unknowns of interest. For each of these we derive a gradient flow in Appendix B that can be shown to decrease

---

[3]This statement is exactly true only when the depth map is an equifocal plane.

the cost at each iteration. The final form is

$$
\begin{aligned}
\nabla_s E(x) &= -2 \int_0^{t_1} \left(\frac{\hat{W}^1(x)(\hat{W}^1(x))^T}{s(x)} \nabla u_1(x,t) \right) \cdot \nabla w_1(x, t_1 - t)\, dt \\
&\quad - 2 \int_0^{t_1} \left(\frac{\hat{W}^2(x)(\hat{W}^2)^T(x)}{s(x)} \nabla u_2(x,t) \right) \cdot \nabla w_2(x, t_1 - t)\, dt \\
&\quad - 2\alpha_0 \Delta s(x) + 2\alpha_1 \left(\int_\Omega s(y) dy - s^{mean} \right) \\
&\quad - 2\alpha_2 \frac{\|\hat{W}^1(x)\|^2 + \|\hat{W}^2(x)\|^2}{s(x)} \\
\nabla_{V^1} E &= -2 \int_0^{t_1} \int_\Omega \left(B_1^1(x) \nabla u_1(x,t) \right) \cdot \nabla w_1(x, t_1 - t)\, dx\, dt \\
&\quad - 2 \int_0^{t_1} \int_\Omega \left(B_1^2(x) \nabla u_2(x,t) \right) \cdot \nabla w_2(x, t_1 - t)\, dx\, dt \\
&\quad + 2\alpha_2 \int_\Omega \hat{W}^1(x) \hat{w}_{V^1}^1(x) + \hat{W}^2(x) \hat{w}_{V^1}^2(x) dx \\
\nabla_{V^2} E &= -2 \int_0^{t_1} \int_\Omega \left(B_2^1(x) \nabla u_1(x,t) \right) \cdot \nabla w_1(x, t_1 - t)\, dx\, dt \\
&\quad - 2 \int_0^{t_1} \int_\Omega \left(B_2^2(x) \nabla u_2(x,t) \right) \cdot \nabla w_2(x, t_1 - t)\, dx\, dt \\
&\quad + 2\alpha_2 \int_\Omega \hat{W}^1(x) \hat{w}_{V^2}^1(x) + \hat{W}^2(x) \hat{w}_{V^2}^2(x) dx \\
\nabla_b E &= -2 \int_0^{t_1} \int_\Omega \left(F(x) \nabla u_1(x,t) \right) \cdot \nabla w_1(x, t_1 - t)\, dx\, dt \\
&\quad - 2 \int_0^{t_1} \int_\Omega \left(F(x) \nabla u_2(x,t) \right) \cdot \nabla w_2(x, t_1 - t)\, dx\, dt \\
&\quad + 2\alpha_2 \int_\Omega \frac{(\hat{W}^1(x))^T \hat{W}^2(x) + (\hat{W}^2(x))^T \hat{W}^1(x)}{\sqrt{b^2 + 1}} dx,
\end{aligned}
$$

(8.47)

where the adjoint solution w_j satisfies

$$
\begin{cases}
\dot{w}_j(x,t) = \nabla \cdot \left(\hat{D}_j(x) \nabla w_i(x,t) \right) \\
w_j(x,0) = \begin{cases} u_1(x, t_1) - u_2(x, t_1) & \text{if } j = 1 \\ u_2(x, t_1) - u_1(x, t_1) & \text{if } j = 2 \end{cases} \\
(\hat{D}_j(x) \nabla w_j(x,t)) \cdot n(x) = 0
\end{cases}
$$

(8.48)

and for $i = 1, 2$ and $j = 1, 2$,

$$
\begin{aligned}
B_i^j(x) &= \frac{v}{s(x)} \left(\hat{W}^j(x) \hat{w}_{V^i}^j(x) [1\ 1] + \begin{bmatrix} 1 \\ 1 \end{bmatrix} \left(\hat{W}^j(x) \hat{w}_{V^i}^j(x) \right)^T \right) \\
F(x) &= \frac{1}{\sqrt{b^2 + 1}} \left(\hat{W}^1(x)(\hat{W}^2(x))^T + \hat{W}^2(x)(\hat{W}^1(x))^T \right)
\end{aligned}
$$

(8.49)

Table 8.2. Summary of the depth map and velocities estimation algorithm with motion blur along different directions.

Algorithm (motion blur along different directions)

1. Given: Calibration parameters (from knowledge of the camera; see Appendix D for a simple calibration procedure) $v, F, D, \gamma, \Delta\tilde{T}_1, \Delta\tilde{T}_2$, two images $J(x, \Delta\tilde{T}_1)$, $J(x, \Delta\tilde{T}_2)$, a chosen threshold ϵ, regularization parameters $\alpha_0, \alpha_1, \alpha_2$, the scale parameter s^{mean}, and step size $\beta_s, \beta_{V^1}, \beta_{V^2}$, and β_b seek the depth map $\hat{s}$, motion vectors $\hat{V}^1_{1,2}$, $\hat{V}^2_{1,2}$, and parameter $\hat{b}$ as follows.

2. Initialize depth map $\hat{s}(x, 0)$ with a plane at depth
$$\hat{s}(x, 0) = \frac{(v_1 + v_2)F}{v_1 + v_2 - 2F};$$
initialize velocities $\hat{V}^1_{1,2}(0)$, $\hat{V}^2_{1,2}(0)$ to two orthonormal vectors with random initial direction; initialize $\hat{b}(0) = 0$.

3. Compute the diffusion tensors $\hat{D}_j$ via equation (8.28) and equations (8.31) and (8.32),

4. Simulate (i.e. numerically evaluate) equations (8.26) and (8.27),

5. Using the solution obtained at the previous step, simulate equations (8.48),

6. Compute the gradient of u and w and evaluate equations (8.47),

7. Update the unknowns via
$$
\begin{aligned}
\hat{s}(x, k+1) &= \hat{s}(x, k) - \beta_s \nabla_s E(x) \\
\hat{V}^1_{1,2}(k+1) &= \hat{V}^1_{1,2}(k) - \beta_{V^1} \nabla_{V^1} E \\
\hat{V}^2_{1,2}(k+1) &= \hat{V}^2_{1,2}(k) - \beta_{V^2} \nabla_{V^2} E \\
\hat{b}(k+1) &= \hat{b}(k) - \beta_b \nabla_b E.
\end{aligned}
$$

8. Return to Step 3 until norm of gradients is below the chosen threshold ϵ.

and
$$
\begin{aligned}
\hat{w}^1_{V^1}(x) &\doteq \frac{v}{s(x)}\sqrt{b^2 + 1}\,\Delta\tilde{T}_1 \\
\hat{w}^2_{V^1}(x) &\doteq \frac{v}{s(x)}b\Delta\tilde{T}_1 \\
\hat{w}^1_{V^2}(x) &\doteq \frac{v}{s(x)}b\Delta\tilde{T}_2 \\
\hat{w}^2_{V^2}(x) &\doteq \frac{v}{s(x)}\sqrt{b^2 + 1}\,\Delta\tilde{T}_2.
\end{aligned}
\tag{8.50}
$$

We summarize the algorithm in Table 8.2.

8.5   Image restoration

Scene structure, motion, and image restoration are different aspects of the same problem. So far we have chosen to focus on the first two, and we have therefore employed the notion of relative blur to eliminate the unknown radiance. However, in many applications one is interested precisely in the radiance, that can be thought of as a deblurred image. In this section, therefore, we increase the generality of the estimation problem to include image restoration as well as motion estimation, while still allowing for multiple objects with different motion. However, to avoid the ambiguities in the reconstruction of the velocity field that we analyzed in the previous section, that forced us to introduce the artificial parameter b (see equations (8.31) and (8.32)), we use $K > 2$ images as input. We assume that the velocity of one object in all the images is never along the same direction. This is sufficient to constrain the problem so that there are at most two isolated solutions for the velocity vectors. In the case of velocities along the same direction the solution is unique, as we impose that the shutter intervals of all the images are different. In the following we analyze the solution in the more difficult case of an object that is always changing motion direction.

Suppose we are given $K = 3$ images and assume that there is a single moving object in the scene. Let us denote with $V_{1,2}^i$, $i = 1, 2, 3$ the velocities of the object in images $J(\cdot, \Delta\tilde{T}_1)$, $J(\cdot, \Delta\tilde{T}_2)$, and $J(\cdot, \Delta\tilde{T}_3)$. Then we know from the previous section that the estimated velocities $\hat{W}^1$, $\hat{W}^2$, $\hat{W}^3$ must satisfy

$$
\begin{aligned}
\hat{W}^1(x) &= \sqrt{b^2 + 1}\, v\Delta\tilde{T}_1 \frac{V_{1,2}^1}{s(x)} + bv\Delta\tilde{T}_2 \frac{V_{1,2}^2}{s(x)} \\
\hat{W}^2(x) &= bv\Delta\tilde{T}_1 \frac{V_{1,2}^1}{s(x)} + \sqrt{b^2 + 1}\, v\Delta\tilde{T}_2 \frac{V_{1,2}^2}{s(x)} \\
\hat{W}^1(x) &= \sqrt{c^2 + 1}\, v\Delta\tilde{T}_1 \frac{V_{1,2}^1}{s(x)} + cv\Delta\tilde{T}_3 \frac{V_{1,2}^3}{s(x)} \\
\hat{W}^3(x) &= cv\Delta\tilde{T}_1 \frac{V_{1,2}^1}{s(x)} + \sqrt{c^2 + 1}\, v\Delta\tilde{T}_3 \frac{V_{1,2}^3}{s(x)}
\end{aligned}
\tag{8.51}
$$

for scalars $b, c \in \mathbb{R}$. By equating the solutions in $\hat{W}^1$, we obtain that

$$
\left(\sqrt{b^2 + 1} - \sqrt{c^2 + 1}\right)\Delta\tilde{T}_1 V_{1,2}^1 = c\Delta\tilde{T}_3 V_{1,2}^3 - b\Delta\tilde{T}_2 V_{1,2}^2.
\tag{8.52}
$$

Then, let us write the velocity $V_{1,2}^3$ as a linear combination of the other two velocity vectors; that is,

$$
V_{1,2}^3 = \nu_1 \frac{\Delta\tilde{T}_1}{\Delta\tilde{T}_3} V_{1,2}^1 + \nu_2 \frac{\Delta\tilde{T}_2}{\Delta\tilde{T}_3} V_{1,2}^2
\tag{8.53}
$$

for some scalars $\nu_1, \nu_2 \in \mathbb{R}$. By multiplying equation (8.52) on both sides by a vector orthogonal to $V_{1,2}^1$ and by substituting the explicit expression of $V_{1,2}^3$ above, we obtain

$$
c = \frac{b}{\nu_2};
\tag{8.54}
$$

by multiplying equation (8.52) on both sides by a vector orthogonal to $V_{1,2}^2$ and by substituting the explicit expression of $V_{1,2}^3$ and of c above, we obtain

$$\sqrt{b^2 + 1} - \sqrt{\frac{b^2}{\nu_2^2} + 1} = \frac{\nu_1}{\nu_2} b. \tag{8.55}$$

This equation has at most two real solutions in b. One solution is the trivial case of $b = 0$, which corresponds to the desired solution; the other solution exists if the following holds

$$\left(\nu_2^2 - 1 + \nu_1^2\right)^2 > 4\nu_1^2 \nu_2^2 \tag{8.56}$$

and can be explicitly written as

$$b^* = \text{sign}\left(|\nu_2| - 1\right) \frac{2\nu_1 \nu_2}{\sqrt{\left(\nu_2^2 - 1 + \nu_1^2\right)^2 - 4\nu_1^2 \nu_2^2}}. \tag{8.57}$$

Hence, three images are sufficient to constrain the problem of recovering the velocities of the moving objects so that the solution set is finite. In this way we avoid adding artificial constraints to the cost functional as we did in the previous section.

Let us start by introducing the image formation model of a defocused and motion-blurred image $I(x, \Delta \tilde{T}_i)$ via the diffusion model, that we already formulated in the previous chapter. We denote the velocity of the jth object in the ith image with the somewhat cumbersome notation $V_{1,2}^{i,j}$, for which we have already apologized. Then, for $j = 1, \dots, Y$ we have:

$$\begin{cases} \dot{u}_{i,j}(x,t) = \nabla \cdot (D_{i,j}(x) \nabla u_{i,j}(x,t)) & \forall t \in (0, \infty), \forall x \in \Omega_j \\ u_{i,j}(x,0) = r(x) & \forall x \in \Omega_j \\ (D_{i,j}(x) \nabla u_{i,j}(x,t)) \cdot n_j(x) = 0 & \forall x \in \partial \Omega_j \end{cases} \tag{8.58}$$

with

$$D_{i,j}(x) = \sigma^2(x) I_d + v^2 \Delta \tilde{T}_i^2 \frac{V_{1,2}^{i,j} \left(V_{1,2}^{i,j}\right)^T}{s^2(x)} \qquad \forall x \in \Omega_j, \tag{8.59}$$

and $n_j(x)$ with $x \in \partial \Omega_j$ denotes the unit vector normal to the boundary of Ω_j.

Assume that we are given $K > 2$ images $J(\cdot, \Delta \tilde{T}_1), \dots, J(\cdot, \Delta \tilde{T}_K)$ collected while the shutter remains open for different time intervals $\Delta \tilde{T}_1, \dots, \Delta \tilde{T}_K$ and $\Delta \tilde{T}_{i+1} > \Delta \tilde{T}_i$, $i = 1, \dots, K$. Then, one can pose the problem of inferring velocities $\{\hat{V}_{1,2}^{i,j}\}_{j=1,\dots,Y, i=1,\dots,K}$, partitions $\{\hat{\Omega}_j\}_{j=1\dots Y}$, radiance $\hat{r}$, and depth $\hat{s}$ of the scene as the following minimization,

$$\{\hat{V}_{1,2}^{i,j}, \hat{\Omega}_j\}_{i=1,\dots,K, j=1,\dots,Y}, \hat{r}, \hat{s} = \arg \min_{\{V_{1,2}^{i,j}, \Omega_j\}_{i=1,\dots,K, j=1,\dots,Y}, r, s} E, \tag{8.60}$$

where

$$
\begin{aligned}
E \;=\; & \sum_{i=1}^{K}\sum_{j=1}^{Y}\int_{\Omega_j}(u_{i,j}(x,t_1)-J(x,\Delta\tilde{T}_i))^2 dx \\
& + \alpha_0 \left\| r - r^* \right\|^2 \\
& + \alpha_1 \left\| \nabla s \right\|^2 \\
& + \alpha_2 \left(\int_\Omega s(x)dx - s^{mean} \right)^2 \\
& + \alpha_3 \sum_{j=1}^{Y}\int_{\partial\Omega_j} dx
\end{aligned}
\tag{8.61}
$$

and most terms appeared already in equation (7.13) in the previous chapter, except for the last term, which imposes a length constraint on the boundary of each region Ω_j thus penalizing boundaries that are too fragmented or irregular. More details on how to minimize the above cost functional are given in the next section.

8.5.1 Minimization algorithm

To minimize the cost functional above, we follow the same method as in the previous sections. Furthermore, we assume that $Y = 2$ (i.e., there are at most two objects moving in different directions), and have that $\{\hat{r}, \hat{s}, \hat{V}_{1,2}^{i,1}, \hat{V}_{1,2}^{i,2}, \hat{\phi}\}$, $\forall i = 1,\ldots,K$, are the unknowns of interest. The final form of the gradients in this case is

$$
\begin{aligned}
\nabla_r E(x) \;=\;& 2\sum_{i=1}^{K}\Big(H(\phi(x))w_{i,1}(x,t_1) + H(-\phi(x))w_{i,2}(x,t_1)\Big) \\
& + 2\alpha_0(r(x)-r^*(x)) \\
\nabla_s E(x) \;=\;& -2\sum_{i=1}^{K}\int_0^{t_1}(B_{i,1}(x)\nabla u_{i,1}(x,t))\cdot\nabla w_{i,1}(x,t_1-t)\,dt \\
& -2\sum_{i=1}^{K}\int_0^{t_1}(B_{i,2}(x)\nabla u_{i,2}(x,t))\cdot\nabla w_{i,2}(x,t_1-t)\,dt \\
& -2\alpha_1\Delta s(x) + 2\alpha_2\left(\int_\Omega s(y)dy - s^{mean}\right) \\
\nabla_{V^{i,1}} E \;=\;& -2\int_0^{t_1}\int(C_{i,1}(x)\nabla u_{i,1}(x,t))\cdot\nabla w_{i,1}(x,t_1-t)\,dx\,dt \\
\nabla_{V^{i,2}} E \;=\;& -2\int_0^{t_1}\int(C_{i,2}(x)\nabla u_{i,2}(x,t))\cdot\nabla w_{i,2}(x,t_1-t)\,dx\,dt \\
\nabla_\phi E(x) \;=\;& \delta(\phi(x))\left(\sum_{i=1}^{K}(F_{i,1}(x)-F_{i,2}(x)) - 2\alpha_3\nabla\cdot\left(\frac{\nabla\phi(x)}{\|\nabla\phi(x)\|}\right)\right),
\end{aligned}
\tag{8.62}
$$

where

$$
\begin{aligned}
B_{i,1}(x) &= 2H(\phi(x))\left(\frac{\sigma^2(x)}{s^2(x)\left(\frac{1}{F}-\frac{1}{v}\right)-s(x)}I_d - \frac{v^2\Delta\tilde{T}_i^2}{s^3(x)}V_{1,2}^{i,1}(V_{1,2}^{i,1})^T\right) \\
B_{i,2}(x) &= 2H(-\phi(x))\left(\frac{\sigma^2(x)}{s^2(x)\left(\frac{1}{F}-\frac{1}{v}\right)-s(x)}I_d - \frac{v^2\Delta\tilde{T}_i^2}{s^3(x)}V_{1,2}^{i,2}(V_{1,2}^{i,2})^T\right) \\
C_{i,1}(x) &= 2H(\phi(x))\frac{v^2\Delta\tilde{T}_i^2}{s^2(x)}\left(\begin{bmatrix}1\\1\end{bmatrix}(V_{1,2}^{i,1})^T + V_{1,2}^{i,1}[1\ 1]\right) \\
C_{i,2}(x) &= 2H(-\phi(x))\frac{v^2\Delta\tilde{T}_i^2}{s^2(x)}\left(\begin{bmatrix}1\\1\end{bmatrix}(V_{1,2}^{i,2})^T + V_{1,2}^{i,2}[1\ 1]\right) \\
F_{i,1}(x) &= (u_{i,1}(x,t_1) - J(x,\Delta\tilde{T}_1))^2 \\
F_{i,2}(x) &= (u_{i,2}(x,t_1) - J(x,\Delta\tilde{T}_2))^2
\end{aligned}
$$

$$(8.63)$$

and $w_{i,j}$ satisfies the following adjoint system

$$
\begin{cases}
\dot{w}_{i,j}(x,t) = \nabla \cdot (D_{i,j}(x)\nabla w_{i,j}(x,t)) \\
w_{i,j}(x,0) = u_{i,j}(x,t_1) - J(x,\Delta\tilde{T}_i) \\
(D_{i,j}(x)\nabla w_{i,j}(x,t)) \cdot n_j(x) = 0 \qquad \forall x \in \partial\Omega_j.
\end{cases}
$$

$$(8.64)$$

We summarize the algorithm in Table 8.3.

8.6   Examples

In the previous sections we have introduced three algorithms for estimating various parameters of the scene from motion-blurred images. The first two algorithms are designed so as to avoid image restoration, that can optionally be performed in a second step. In particular, the first method considers multiple objects moving with constant velocity along different directions, and the second considers a single object changing direction at each frame. The third method is instead designed for the most complicated case where one performs the simultaneous estimation of all the parameters and there are multiple objects moving along different directions at each image frame. Now we test these algorithms on both real and synthetic data. However, here we are not even attempting to obtain a thorough evaluation of these methods. Rather, we show some results on various aspects of each algorithm to point out their strengths and weaknesses and to provide a guideline to the reader interested in exploring these methods in more detail.

8.6.1   Synthetic data

We start by showing experiments in the case of the algorithm summarized in Table 8.1. We synthetically generate a scene containing two objects; one is made of two circles moving sideways and shown in the left image in Figure 8.2, whereas the other object is the remaining region of the image domain and it is moving along the top-left to bottom-right diagonal. Both objects lie on the same depth

Table 8.3. Summary of the depth map and velocities estimation and image restoration algorithm with motion blur along different directions.

Algorithm (motion blur along different directions)

1. Given: Calibration parameters (from knowledge of the camera; see Appendix D for a simple calibration procedure) $v, F, D, \gamma, \{\Delta \tilde{T}_i\}_{i=1,\ldots,K}$, K images $\{J(x, \Delta \tilde{T}_i)\}_{i=1,\ldots,K}$, a chosen threshold ϵ, regularization parameters α_0, α_1, α_2, and α_3, the scale parameter s^{mean}, and step size β_r, β_s, $\beta_{V^{i,1}}$, $\beta_{V^{i,2}}$, and β_ϕ seek the restored image $\hat{r}$, the depth map $\hat{s}$, motion vectors $\hat{V}_{1,2}^{i,1}$, $\hat{V}_{1,2}^{i,2}$, and level set $\hat{\phi}$ as follows.

2. Initialize the radiance $\hat{r}(x, 0)$ to one of the input images; initialize depth map $\hat{s}(x, 0)$ with a plane at depth

$$\hat{s}(x, 0) = \frac{vF}{v - F};$$

initialize velocities $\hat{V}_{1,2}^{i,1}(0)$, $\hat{V}_{1,2}^{i,2}(0)$ to two orthonormal vectors with random initial direction for all $i = 1, \ldots, K$; initialize level set $\hat{\phi}(x, 0)$ so that it is the signed distance function of uniformly distributed circles on the image domain,

3. Compute the diffusion tensors $\hat{D}_{i,j}$ via equation (8.59),

4. Simulate (i.e., numerically evaluate) equation (8.58),

5. Using the solution obtained at the previous step, simulate equation (8.64),

6. Compute the gradient of u and w and evaluate equations (8.62),

7. Update the unknowns via

$$\begin{aligned}
\hat{r}(x, k+1) &= \hat{r}(x, k) - \beta_r \nabla_r E(x) \\
\hat{s}(x, k+1) &= \hat{s}(x, k) - \beta_s \nabla_s E(x) \\
\hat{V}_{1,2}^{i,1}(k+1) &= \hat{V}_{1,2}^{i,1}(k) - \beta_{V^{i,1}} \nabla_{V^{i,1}} E \\
\hat{V}_{1,2}^{i,2}(k+1) &= \hat{V}_{1,2}^{i,2}(k) - \beta_{V^{i,2}} \nabla_{V^{i,2}} E \\
\hat{\phi}(x, k+1) &= \hat{\phi}(x, k) - \beta_\phi \nabla_\phi E(x).
\end{aligned}$$

8. Return to Step 3 until norm of gradients is below the chosen threshold ϵ.

map, which is the slanted plane shown in Figure 8.2. In the center and right image of Figure 8.2 we also show the reconstructed scene (center image) with the estimated depth map (right image). The estimated depth map is shown as a gray-level image. Light intensities correspond to points that are close to the viewer, and dark intensities correspond to points that are far from the viewer. When the scene is static, the image we capture coincides with the radiance of the object (leftmost image in Figure 8.3). The second and third images from the left of Figure 8.3

Figure 8.2. Left: Setup of the scene with motion blur. The depth map is a slanted plane. The steps on the top are closer to the camera than the steps on the bottom. Two disks on the plane shift from left to right, and the remaining part of the plane shifts along the top-left to bottom-right diagonal. The texture of the two disks has been brightened to make them more visible. Center: Reconstruction of the setup of the scene by using the recovered radiance and the reconstructed depth map. Right: Visualization of the estimated depth map as a gray-level image. Light intensities correspond to points that are close to the camera, and dark intensities correspond to points that are far from the camera.

Figure 8.3. First from the left: Synthetically generated radiance. Second and third from the left: Motion-blurred images captured with different shutter intervals. The motion blur of the third image is three times the motion blur of the second image. Rightmost: Estimated radiance from the two input images. The reconstruction has artifacts at locations corresponding to the boundaries of the disks.

show the two input images captured for different shutter intervals. The shutter interval of the third image is three times the shutter interval of the second image. Also notice that the amount of motion blur is larger on the top of the image than on the bottom. This effect is due to the depth map of the scene. The rightmost image of Figure 8.3 is the resulting deblurred image that we restored from the given input in a second phase by using the image restoration algorithm presented in Section 8.5 with fixed velocities for each object. Notice that the reconstruction is fairly close to the original radiance (leftmost image in Figure 8.2), although there are artifacts at locations corresponding to the boundary of the two disks. This is due to the error between the correct segmentation of the scene and the estimated segmentation (Figure 8.4). In Figure 8.4, top row, we show a few snapshots of the

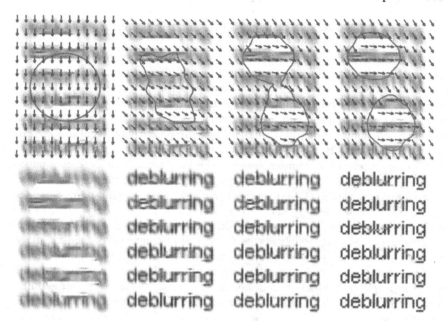

Figure 8.4. Top row: Snapshots of the evolution of segmentation together with motion estimation on synthetic data. Motion is initialized with vertical direction. Bottom row: Snapshots of the evolution of the deblurring of the radiance. The radiance is initialized with the most blurred image (leftmost image). At the second and third iterations, the radiance sharpness improves dramatically (second and third images from the left). The recovered radiance (rightmost image) compares well with the original radiance (leftmost image in Figure 8.3).

segmentation evolution of the two moving objects. The motion field direction is correctly estimated. Also, notice that the level set representation easily handles topological changes of the represented contour. In Figure 8.4, bottom row, we show some snapshots of the deblurring evolution. More precisely, the first three snapshots from the left correspond to the first three steps in the iterative scheme, and the rightmost snapshot corresponds to the last estimation step of the radiance. Although the radiance is initialized with the most blurred image (third image from the left in Figure 8.3), it converges rather quickly to the deblurred image.

8.6.2   Real data

The simple test in the previous section served the purpose of giving a rough idea of what to expect when testing real images. Indeed, in the case of real images there is a clear discrepancy between the proposed model and the true physical process as we pointed out earlier on. The major discrepancy is at occlusion boundaries, where the scene should be represented by two overlapping surfaces, and it is instead represented by one. We examine that case in the next chapter and show how to explicitly model such a phenomenon.

Before showing the results, it is necessary to spend a few words on how to capture real images with different shutter intervals. Although we already discussed this matter in Remark 2.3, we report it here again for ease of reading. The most conceptually straightforward method is to use different cameras that can simultaneously capture images with different shutter intervals. However, in this case one might encounter some difficulties in registering the images and in synchronizing the cameras. [Ben-Ezra and Nayar, 2003] describes hardware that can be used in this modality. A simpler way to capture images with different shutter intervals is to collect a sequence of images. Time averaging the sequence simulates a long shutter interval. In our case we collected three motion-blurred images $[\bar{J}_1, \bar{J}_2, \bar{J}_3]$, and then let $J(\cdot, \Delta\tilde{T}_1) = \bar{J}_2$ and $J(\cdot, \Delta\tilde{T}_2) = 1/3 \sum_{i=1}^{3} \bar{J}_i$ as the images given as input to the algorithm. The (virtual) shutter interval for the second image $J(\cdot, \Delta\tilde{T}_2)$ is three times the (virtual) shutter interval of the first image $J(\cdot, \Delta\tilde{T}_1)$; that is, $\Delta\tilde{T}_2 = 3\Delta\tilde{T}_1$. In this case, the data collection is rather simple because no alignment and no synchronization is required, but it is based on the strong assumption that motion does not change among the three frames and that time averaging is a good approximation to large motion blur. Our experiments show that such a modality is a reasonable approximation on the test images. In Figure 8.5 on the top-left corner we show an image of the scene when static. This image approximates the radiance of the scene. On the top-right corner we show the recovered image obtained by using Algorithm 8.3. Although the two images are not perfectly aligned (one can notice a small misalignment between the background of the image to the left and the background of the image to the right of Figure 8.5), the reconstruction resembles the radiance of the scene. Furthermore, as mentioned above, there are artifacts at locations corresponding to the boundary of the segmented regions. As an input we use the image on the bottom-left corner (which corresponds to $J(\cdot, \Delta\tilde{T}_1)$) and the image on the bottom-right corner (which corresponds to $J(\cdot, \Delta\tilde{T}_2)$) of Figure 8.5. The background is moving vertically, whereas the foreground (the cup and the banana) are moving horizontally. In Figure 8.6 we show a few snapshots of the segmentation evolution. To make the contour more visible in the illustrations, we changed the original brightness of the image. Notice that the motion field direction of the scene is correctly estimated. In Figure 8.7 we show some snapshots of the deblurring evolution. We use as initial radiance the most blurred image (left). The final estimate of the radiance (right) is also shown in Figure 8.5 for comparison with the original radiance.

We test Algorithms 8.2 and 8.3 on real data. We consider the more practical scenario, where images are captured sequentially, rather than simultaneously, and the scene is made of objects at different depths (see Figure 8.8). By capturing images sequentially, we give away correspondence. Indeed, in this case one has to also determine the displacement of the same region in different images. This problem is called *tracking* and we do not discuss how this can be carried out in the context of motion blur, as it goes beyond the scope of this book. We therefore

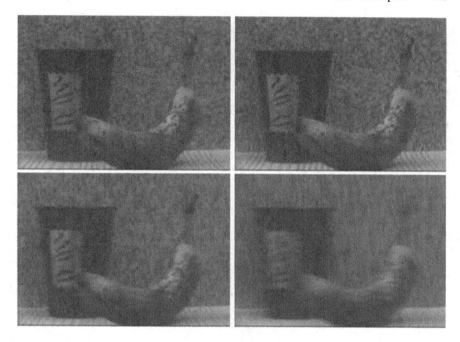

Figure 8.5. Top-left: Original radiance. This image has been captured when the scene and the camera were static. Top-right: Recovered radiance. Bottom-left and bottom-right: Input motion-blurred images. The image on the right has been obtained by averaging three motion-blurred images similar to the image on the left. The shutter interval of the image on the right is approximately three times the shutter interval of the image on the left.

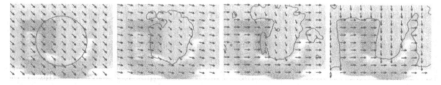

Figure 8.6. Snapshots of the evolution of segmentation. The brightness of one of the motion-blurred images has been changed to enhance the contrast between the image and the contour evolution.

assume that correspondence is given by some tracking algorithm, for instance [Jin et al., 2005].

In this experiment, the camera is moving along a circular trajectory as can be seen in both images in Figure 8.8. This type of motion causes the scene to be segmented into a number of regions, due to the crude approximation of the true velocity field (a rigid motion under perspective projection) with a simple fronto-parallel translational model. However, here we do not show such segmentation, but only analyze the region around the patch marked in Figure 8.8. Notice that in

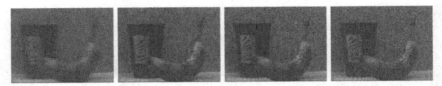

Figure 8.7. Top-left: Original radiance. This image has been captured when the scene and the camera were static. Top-right: Recovered radiance. Bottom-left and bottom-right: Input motion-blurred images. The image on the right has been obtained by averaging three motion-blurred images similar to the image on the left. The shutter interval of the image on the right is approximately three times the shutter interval of the image on the left.

Figure 8.8. Left: Trajectory of the tracked image region superimposed onto the first frame. The arrows indicate the estimated motion blur directions and intensities at each frame. Notice that, due to the rapid change of direction, the motion-blur velocity vectors differ from the displacement vector between adjacent frames. Right: Trajectory of the tracked image region superimposed to the last frame of the original sequence.

this experiment each frame is motion blurred along different directions. On the left image we superimpose the given image region (a 25×25 pixel patch) trajectory together with the estimated motion-blur velocity vectors. The motion-blur velocity vectors are approximately tangential to the trajectory and are in general different from the relative displacements between adjacent frames. On the right image we show the last frame of the sequence. In Figure 8.9 we test Algorithm 8.2 on pairs of images made of the first frame and the ith frame. We show how the selected image region in the first frame is blurred to match the region in all subsequent frames, and how the corresponding regions in all subsequent frames are blurred to match the region in the first frame. We denote with $\hat{W}^{1,i}$ the estimated velocity of the first frame in the ith pair, and with $\hat{W}^{2,i}$ the estimated velocity of the ith frame in the ith pair. This can be understood more clearly by looking at Figure 8.9. The top row shows a collection of patches extracted from the original images. The middle row shows the patch from the first frame (corresponding to the leftmost image of the top row) motion blurred with velocities $\hat{W}^{2,i}$, $i = 1, \ldots, 9$, so as to match the corresponding patches in the bottom row. The bottom row shows the patch in the ith image, $i = 1, \ldots, 9$ motion blurred with velocity $\hat{W}^{1,i}$.

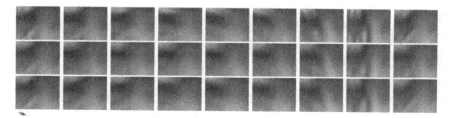

Figure 8.9. Top row: Snapshots from the original sequence (undistorted). Middle row: Snapshots of the image region from the first frame motion blurred so as to match the corresponding image region at subsequent frames. Bottom row: Snapshots of the image region at subsequent frames motion blurred and deformed so as to match the image region at the first frame.

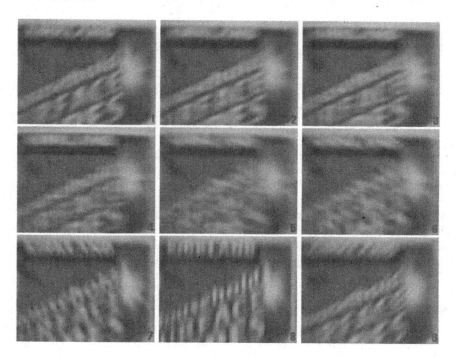

Figure 8.10. Snapshots of large image regions from the original sequence. Due to motion blur, most of the text is unreadable.

Finally, we test Algorithm 8.3 on image restoration. We use the motion vectors estimated with Algorithm 8.2 to initialize the motion vectors in Algorithm 8.3. Then, we show the performance of Algorithm 8.3 in a larger region (see Figure 8.10). We use the nine frames employed in Algorithm 8.2 and simultaneously estimate motion, and shape and perform image restoration. For simplicity, in this experiment we assume that the depth map in this region is constant. In Figure 8.11

Figure 8.11. Snapshots from the image restoration evolution. As one can see, the final deblurred image (bottom row-rightmost) reveals details that were not visible on any of the original image regions (Figure 8.10).

we show the result of our deblurring procedure. We display only nine of the total thirty iterations (see number at the bottom-right corner of each image in Figure 8.11). Notice that the deblurred image in Figure 8.11 reveals details that were not visible on any of the original image regions (see Figure 8.10).

8.7   Summary

In this chapter we have extended the image formation model by allowing multiple objects in the scene to move independently during the shutter interval. The amount of motion blur that they generate conveys information about their motion as well as their shape. However, there are ambiguities in the reconstruction of their 3-D shape, motion, and reflectance that have to be taken into account. We have derived algorithms that extend those of the previous chapter, and their implementation is reported in Appendix E.

9
Dealing with occlusions

So far we have considered scenes where there is no occlusion of line of sight, so that the entire scene is visible from the image, and its shape can therefore be represented by the graph of a function with domain on the image plane. Most often, however, real scenes exhibit complex surfaces that occlude one another. For instance, a pole in front of a wall occludes part of it, and the scene (pole plus wall) cannot be represented by the graph of a function. At first this seems to be a technicality. Because we cannot see beyond the occluding object, we might as well assume that it is connected to the occluder at the occluding boundary, so it can indeed be represented by the graph of a function, albeit not a continuous one. Right? Wrong. This reasoning would be correct if we had a pinhole imaging model, but for a finite-aperture camera, one can actually see portions of the image beyond an occlusion.

To illustrate this fact the reader can perform a simple experiment with a standard photographic camera. As in Figure 2.3, take a background scene (e.g., a picture of a mountain range with some text pasted on it). Then take a foreground scene that partially occludes the background, for instance, a grid as in the jail scene depicted in Figure 2.3. Now, take a photograph with a camera with a very small aperture, akin to a pinhole camera. Cameras with a narrow aperture have a small depth of field, so both the background and the foreground appear in focus and, as one would expect, the background is only partially visible because of the occlusion. Finally, take a photograph with the same camera but now with a wide aperture, bringing the background into focus. As one can see in Figure 2.3, one can resolve the entire background scene, as if the foreground were "blurred away." The extent of this phenomenon, which is well known and exploited in confocal microscopy, depends on the aperture, focal length, size of the occluding object, and relative distance from the background.

In this chapter we show how to extend the algorithms of previous chapters to handle occlusions. Naturally we have to enrich the representation, because surfaces in the scene can no longer be modeled as the graph of a function. The model we use for the occlusion is the one introduced in Section 2.3. We devise an algorithm to infer both the shape of the scene, the location of the occluding boundary, and its radiance, including the portions that are not visible when imaged through a pinhole.

9.1   Inferring shape and radiance of occluded surfaces

Recall the notation presented in Section 2.3. We denoted the surface and the radiance of the occluding object (Object 2) with s_2 and r_2, respectively, and the surface and the radiance of the occluded object (Object 1) were denoted by s_1 and r_1, respectively. The spatial support of the occluding object is defined by the function $\phi : \mathbb{R}^2 \mapsto \mathbb{R}$ via equation (2.24). The imaging model, given in equation (2.27) for a certain focus setting v, is

$$
\begin{aligned}
I(y, v) = & \int_\Omega h^v_{s_2}(y, x) H\big(\phi(x)\big) r_2(x) dx \\
& + \int_\Omega h^v_{s_2}(y, x) \bar{H}\big(\phi(\varpi(x))\big) r_1(x) dx,
\end{aligned}
\tag{9.1}
$$

where ϖ is the projection of points on the surface of Object 2 through the point A of coordinates $(-(yu_0)/v, u_0)$ on the surface of Object 1 (see Figure 2.4). H denotes the Heaviside function (see Section 2.3). Recall also that $J(\cdot, v_i)$ is the observed image corresponding to the imaging model $I(\cdot, v_i)$, for $v = [v_1, \ldots, v_K]$, where K is the total number of images.

In summary, the scene is described by the depth maps s_1 and s_2, the radiances r_1 and r_2, and the support of the occluding object, represented via the function ϕ. All of these functions are unknown and we are given only the measured images $J(\cdot, v_i)$. Our problem is then to infer these unknowns by minimizing a discrepancy measure between the observed image $J(\cdot, v_i)$ and the estimated image $I(\cdot, v_i)$ for each $i = 1 \ldots K$. For instance, as discussed in Section 3.6, we may minimize a quadratic cost functional:

$$
E = \sum_{i=1}^{K} \int_\Omega \left(I(y, v_i) - J(y, v_i) \right)^2 dy.
\tag{9.2}
$$

Since this problem is ill-posed, we need to add regularizing terms, as we discuss in Section 3.6 and Appendix F. For instance, we can add the terms

$$
\alpha_1 \int_\Omega \|r_1(x) - r_1^*(x)\|^2 dx + \alpha_2 \int_\Omega \|r_2(x) - r_2^*(x)\|^2 dx
\tag{9.3}
$$

or some other functional norm, where α_1 and α_2 are positive constants, and r_1^* and r_2^* are the "true" radiances. Because in practice we do not know the true radiances, we can set r_1^* and r_2^* to be one of the input images, or a combination of them, with very small α_1 and α_2. This seems like a hack, and one would rather do without it. However, regularization is necessary in order to make the problem well-posed, and generic regularizers, for instance, functionals that favor smoothness, are not appropriate because the radiance is not smooth. Indeed, we want the radiance to be far from smooth because it corresponds to the ideal deblurred image, with sharp contrast and boundaries. Ideally, one would have a prior model of the radiance of natural scenes, and substitute the above functional with $-\alpha_1 P(r_1) - \alpha_2 P(r_2)$, where P is a probability inferred from collections of deblurred images. Because most computer vision studies assume that images are sharp, one could use the statistics of natural images, rather than those of natural scene radiances, in place of P, as discussed, for instance, in [Huang and Mumford, 1999].

We also need to add regularization to the depth maps. Although depth is also not smooth, because of occluding boundaries, recall that here we are modeling occluding boundaries explicitly, and therefore we can safely assume that within each "layer" the depth map is smooth. Hence we can add to the cost functional a generic regularizing term such as the norm of the gradient of the depth map

$$\alpha_3 \int_\Omega \|\nabla s_1(x)\|^2 dx + \alpha_4 \int_\Omega \|\nabla s_2(x)\|^2 dx \qquad (9.4)$$

for some choice of functional norm, for instance, L^2, where α_3 and α_4 are positive constants. Finally, we impose that the boundary of the support of the occluding object (Object 2) is smooth by adding the term

$$\alpha_5 \int_\Omega \|\nabla H(\phi(x))\|^2 dx, \qquad (9.5)$$

where α_5 is a positive constant (see the appendix for the meaning of the gradient of a discontinuous function, such as the Heaviside H). This term corresponds to the length of the planar contour determined by the interface between positive and zero values in the Heaviside function. By collecting all the terms into a single formula, we obtain the following cost functional $E(r_1, r_2, s_1, s_2, \phi)$,

$$E = \sum_{i=1}^{K} \int_\Omega \left(I(y, v_i) - J(y, v_i)\right)^2 dy + \alpha_1 \|r_1 - r_1^*\|^2 + \alpha_2 \|r_2 - r_2^*\|^2$$
$$+ \alpha_3 \|\nabla s_1\|^2 + \alpha_4 \|\nabla s_2\|^2 + \alpha_5 \|\nabla H(\phi)\|^2. \qquad (9.6)$$

This type of cost functional is common in blind image restoration, where there is no knowledge of the imaging kernel [You and Kaveh, 1999; You and Kaveh, 1996]. There, the only unknown in the imaging kernel is a scalar parameter at each point, which determines the depth of the corresponding layer.

The five unknowns r_1, r_2, s_1, s_2 and ϕ can be inferred by minimizing the cost functional E; that is,

$$\hat{r}_1, \hat{r}_2, \hat{s}_1, \hat{s}_2, \hat{\phi} = \arg \min_{r_1, r_2, s_1, s_2, \phi} E. \tag{9.7}$$

The minimization can be carried out in an iterative fashion. As we have done in the previous chapters, we introduce an artificial time variable k that indicates the iteration number, so that each unknown acquires one more parameter. For example, the estimate of r_2 is a function $\hat{r}_2 : \Omega_2 \times [0, \infty) \mapsto [0, \infty)$, such that, if the solution is unique and convergence is attained, then $\hat{r}_2 \doteq \lim_{k \to \infty} \hat{r}_2(x, k) = r_2(x)$. We call *evolution* of r_2 the change of $\hat{r}_2$ in (iteration) time. Now, let $\xi : \mathbb{R}^2 \times [0, \infty) \mapsto \mathbb{R}$ represent one of the estimated unknowns as it evolves in (iteration) time. Then, to minimize the cost functional (9.6), we use the following scheme

$$\frac{d\xi}{dk} = -\nabla_\xi E, \tag{9.8}$$

where $\nabla_\xi E$ is the first-order variation of the cost functional with respect to ξ and $\nabla_\xi E = 0$ is the corresponding Euler–Lagrange equation, introduced in Section B.1.2. Notice that when the Euler–Lagrange equation is satisfied (i.e., when the necessary conditions for a minimum are satisfied), ξ remains constant. Furthermore, the flow is guaranteed to minimize the cost functional as we have seen in equation (5.13) in Section 5.2. As usual, even when the cost functional is minimized, there is no guarantee that the minimizer, that is, the shape and radiance of the scene, corresponds to the ground truth. For this to happen, observability (uniqueness) conditions would have to be satisfied. For scenes with occlusions this is a difficult problem that does not yet have a satisfactory answer.

On practical grounds, the most crucial issue is that of initialization, which we address in the following sections.

9.2   Detecting occlusions

Despite the fact that the above formulation can model scenes made of a single surface as a degenerate case, there is a considerable increase in computational complexity when compared to the techniques described in previous chapters. It is therefore desirable to use the model for occlusions only where it is necessary. This requires us to identify occluding boundaries at the outset, before applying the minimization scheme outlined above. The simplest method to detect occluding boundaries is to assume that the scene is made of a single smooth surface, and then look at the residual of the cost function. Occluding boundaries elicit a high residual, whereas smooth variations in depth do not. Therefore, given a set of blurred images (see Figure 9.2), we can first compute the residual with a single-surface/single-radiance model (see Figure 9.1); then, we can select regions whose corresponding residuals are above a chosen threshold (see Figure 9.1), and restrict the occlusion model to operate within those regions. This step is conceptually not necessary, but it is beneficial in speeding up the computation.

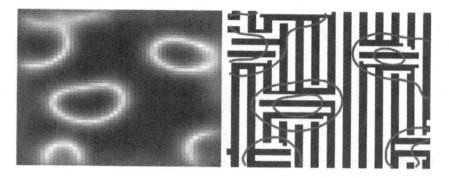

Figure 9.1. Left: residual of the cost function when assuming a single-surface/single-radiance model (white denotes high residual). High residual corresponds to occluding boundaries. Right: a pinhole image of the scene in Figure 9.2 with detected occlusion regions. Smooth variations in depth do not elicit large residuals.

9.3   Implementation of the algorithm

In this section we give some details about the minimization of the cost functional (9.6). Although we encountered most of the terms in the previous chapters, the derivation of the gradient is slightly different from before, because we are employing the convolutional model in equation (9.1) rather than the PDE-based model. The details about these computations can be found in Section B.2.4. Here we only report the final computations of the first-order variations of the cost functional with respect to the five unknowns $\{\bar{r}_1, r_2, s_1, s_2, \phi\}$. The gradients are:

$$\nabla_{\bar{r}_1} E(\bar{x}) = 2\left(1 - H\left(\phi(\bar{x})\right)\right) \sum_{i=1}^{K} \int \left(\bar{I}(\hat{y}, v_i) - \bar{J}(\hat{y}, v_i)\right) \frac{s_1(\hat{y}) - u_0}{s_2(\hat{y}) - u_0}$$

$$h_{s_1(\hat{y})}^{v_i}\left(\hat{y}\frac{s_1(\hat{y})}{s_2(\hat{y})}, \varpi^{-1}(\bar{x})\right) d\hat{y} + 2\alpha_1(\bar{r}_1(\bar{x}) - r_1^*(\bar{x})) \quad (9.9)$$

$$\nabla_{r_2} E(\bar{x}) = 2H\left(\phi(\bar{x})\right) \sum_{i=1}^{K} \int \left(\bar{I}(\hat{y}, v_i) - \bar{J}(\hat{y}, v_i)\right) h_{s_2(\hat{y})}^{v_i}(\hat{y}, \bar{x})\, d\hat{y}$$

$$+ 2\alpha_2(r_2(\bar{x}) - r_2^*(\bar{x})) \quad (9.10)$$

$$\nabla_{s_1} E(\hat{y}) = \frac{E(s_1(\hat{y}) + ds_1(\hat{y})) - E(s_1(\hat{y}))}{ds_1(\hat{y})} \quad (9.11)$$

$$\nabla_{s_2} E(\hat{y}) = \frac{E(s_2(\hat{y}) + ds_2(\hat{y})) - E(s_2(\hat{y}))}{ds_2(\hat{y})} \quad (9.12)$$

$$\nabla_\phi E(\bar{x}) = 2\delta(\phi(\bar{x})) \sum_{i=1}^{K} \int \left(\bar{I}(\hat{y}, v_i) - \bar{J}(\hat{y}, v_i) \right) \left(h^{v_i}_{s_2(\hat{y})}(\hat{y}, \bar{x}) r_2(\bar{x}) \right.$$

$$\left. - h^{v_i}_{s_1(\hat{y})} \left(\hat{y} \frac{s_1(\hat{y})}{s_2(\hat{y})}, \varpi^{-1}(\bar{x}) \right) \bar{r}_1(\bar{x}) \frac{s_1(\hat{y}) - u_0}{s_2(\hat{y}) - u_0} \right) d\hat{y}$$

$$- 2\alpha_5 \delta(\phi(\bar{x})) \nabla \cdot \left(\frac{\nabla \phi(\bar{x})}{|\nabla \phi(\bar{x})|} \right),$$

$$\tag{9.13}$$

where the projected image $\bar{I}(\cdot, v_i)$ satisfies

$$\bar{I}(\hat{y}, v_i) = I\left(-\frac{\hat{y}v}{s_2(\hat{y})}, v_i \right) = I(y, v_i) \tag{9.14}$$

and

$$\bar{r}_1(\bar{x}) = r_1 \left(\varpi^{-1}(\bar{x}) \right). \tag{9.15}$$

The algorithm is summarized in Table 9.1.

9.4 Examples

In this section we show some experiments on real and synthetically generated data. We consider scenes where one object (or a collection of objects) in the foreground occludes another in the background (see Figure 9.2 or Figure 9.6 for examples). For each experiment we use four images: two captured so that the foreground object is approximately in focus and the other two captured so that the background object is approximately in focus. The distances at play are such that the amount of blur on either the foreground object or the background object is large. In our current implementation we use a Gaussian kernel, but the algorithm is not restricted to this choice (see Chapter 2).

9.4.1 Examples on a synthetic scene

The numerical evaluation of the performance of the algorithm is quite complex due to the wide set of scenarios and parameters involved. The major difficulty is in comparing shapes (here represented by two surfaces), which cannot be done as in the previous chapters by simply computing the root mean square error between two functions (the surfaces). In fact, now the surfaces cannot be represented by a single function, and even may not be defined on the same domain of the corresponding true surface.

Away from discontinuities, the performance of the algorithm can be roughly evaluated as in previous chapters. Here, we show some qualitative experiments that serve the purpose of illustrating in a concise fashion the capabilities of the estimation scheme, rather than giving a precise assessment of its performance.

Table 9.1. Summary of the algorithm to recover radiances r_1 and r_2, surfaces s_1, and s_2 and support function ϕ in the presence of occlusions.

Algorithm (occlusions)

1. Given: Calibration parameters (from knowledge of the camera; see Appendix D for a simple calibration procedure) $\{v_i\}_{i=1,\ldots,K}, F, D, \gamma, K$ images $J(\cdot, v_1), \ldots, J(\cdot, v_K)$, a chosen threshold ϵ, regularization parameters $\alpha_1, \alpha_2, \alpha_3, \alpha_4, \alpha_5$, the priors r_1^* and r_2^*, and step sizes $\beta_{\bar{r}_1}, \beta_{r_2}, \beta_{s_1},$ $\beta_{s_2},$ and β_ϕ, seek radiances $\hat{r}_1$ and $\hat{r}_2$, surfaces $\hat{s}_1$ and $\hat{s}_2$, and support $\hat{\phi}$ as follows.

2. Initialize depth map $\hat{s}_1(\hat{y}, 0)$ with a plane at depth such that it is in focus for one of the focus settings and the farthest away from the camera; similarly, initialize depth map $\hat{s}_2(\hat{y}, 0)$ with a plane at depth such that it is in focus for one of the focus settings and the closest to the camera; initialize the radiances $\hat{r}_1(\bar{x}, 0)$ and $\hat{r}_2(\bar{x}, 0)$ with the images corresponding to the focus setting used to set $\hat{s}_1(\hat{y}, 0)$ and $\hat{s}_2(\hat{y}, 0)$, respectively; initialize the support $\hat{\phi}(\bar{x}, 0)$ so that it is the signed distance function of uniformly distributed circles on the image domain.

3. Compute the gradients in equation (9.9) and subsequent equations,

4. Update the unknowns via

$$\hat{r}_1(\bar{x}, k+1) = \hat{r}_1(\bar{x}, k) - \beta_{\bar{r}_1} \nabla_{\bar{r}_1} E(\bar{x})$$
$$\hat{r}_2(\bar{x}, k+1) = \hat{r}_2(\bar{x}, k) - \beta_{r_2} \nabla_{r_2} E(\bar{x})$$
$$\hat{s}_1(\hat{y}, k+1) = \hat{s}(\hat{y}, k) - \beta_{s_1} \nabla_{s_1} E(\hat{y})$$
$$\hat{s}_2(\hat{y}, k+1) = \hat{s}(\hat{y}, k) - \beta_{s_2} \nabla_{s_2} E(\hat{y})$$
$$\hat{\phi}(\bar{x}, k+1) = \hat{\phi}(\bar{x}, k) - \beta_\phi \nabla_\phi E(\bar{x}).$$

5. Return to Step 3 until the norm of the gradients is below the chosen threshold ϵ.

6. Project the radiance $\hat{r}_1$ to the original domain by using equation (9.15).

We generate four images from a scene made of an equifocal plane in the background, and a collection of surfaces in the foreground. The surfaces in the foreground all lie on another equifocal plane, that is, closer to the camera (see Figure 9.2). The images are generated by changing the focus setting, and then the relative scaling is corrected for as explained in Appendix D (see Figure 9.3). We estimate the radiance on the background surface (vertical stripes), and the radiance on the foreground surfaces (horizontal stripes). In Figure 9.3 we show a few snapshots of the estimation process of both the radiances, together with the ground truth (right). In Figure 9.4 we visualize the evolution of the zero level set of the support function ϕ as a contour superimposed on the pinhole image of the

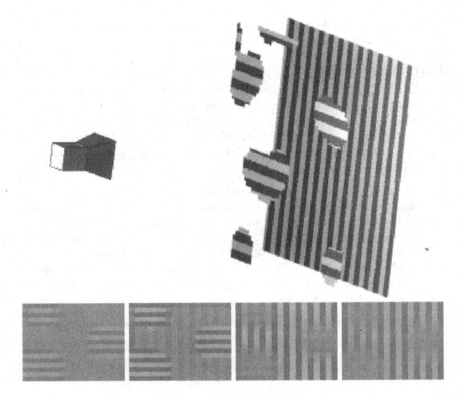

Figure 9.2. Synthetic scene with occlusions. Top: Setup. Bottom: Data. From left to right we bring into focus equifocal planes that are more and more away from the camera; thus, first we bring the foreground object into focus and then the background object.

scene. In Figure 9.5 we show some snapshots from the evolution of the surfaces estimation. In particular, the foreground surfaces are shown only on the estimated support, and both surfaces are texture-mapped with the estimated radiances.

9.4.2 Examples on real images

In Figure 9.6 we show the setup used in our experiments with real scenes. In the background we put the "IEEE" sign over a highly textured surface at a distance of 1.10 m from the camera. In the foreground we placed a set of stripes at a distance of 0.40 m, in such a way as to completely cover the "IEEE" sign. However, when the background is focused, the finite size of the lens allows us to see "behind" the foreground object as evident in Figures 9.6 and 9.7. To capture the images we used a Nikon AF NIKKON 35 mm lens and focus at 0.35 m, 0.45 m, 1.05 m, and 1.15 m. Notice that the setup image in Figure 9.6 has been obtained by placing the camera and the objects closer to each other so that the stripes and the sign "IEEE" are both clearly visible.

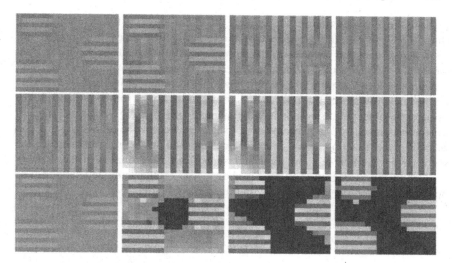

Figure 9.3. Experiments on synthetic data. Top row: Four patches of 15×15 pixels each generated synthetically are shown. Middle row: On the left we show three snapshots of the evolution of r_1, and on the right we show the ground truth for comparison. Bottom row: On the left we show three snapshots of the evolution of r_2 together with the estimated support, and on the right we show the ground truth (on the true support).

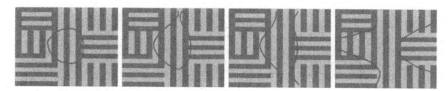

Figure 9.4. Synthetic scene. Evolution of support function ϕ defined in equation (2.24). On the leftmost image we show the initial contour, and on the rightmost image we show the contour at convergence. The zero level set of the support function is shown superimposed on the pinhole image of the scene.

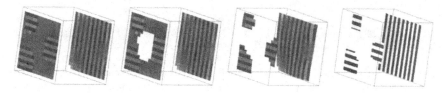

Figure 9.5. Synthetic scene. From the left, three snapshots of the evolution of surfaces s_1 and s_2 together with the support ϕ defined in equation (2.24). The surface s_2 is shown only on the domain estimated via ϕ. On the rightmost image we show the true scene geometry for comparison.

As in the previous section, we estimate the radiance on the background surface
(the "IEEE" sign), and the radiance on the foreground surfaces. In Figure 9.7 we
show a few snapshots of the estimation process of both the radiances. On the right
we put the "ground truth" for comparison. The image for the radiance in the background has been taken by simply removing the occluding object and by focusing
on the background surface. Instead, the ground truth image for the radiance in the
foreground object has been obtained by manually segmenting a focused image of
the foreground object. In Figure 9.8 we visualize the evolution of the zero level set
of the support function ϕ as a contour superimposed on the pinhole image of the
scene. In Figure 9.9 we show some snapshots from the evolution of the surfaces
estimation. The foreground surface is shown only on the estimated support, and
both surfaces are texture-mapped with the estimated radiances.

Figure 9.6. Left: Camera and scene setup. Right: Two of the images captured for two different focal settings. As can be seen, on the top-right image the foreground object completely
occludes the background object. However, by bringing the background in focus, the sign
"IEEE" on the background image becomes visible due to the finite aperture of the lens
(bottom-right image) similarly to Figure 2.3.

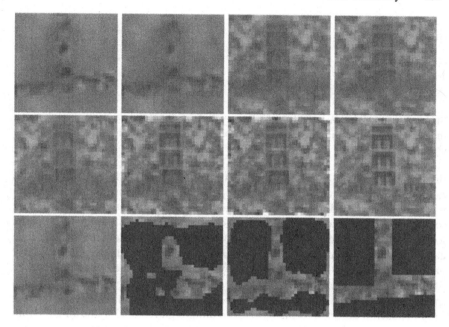

Figure 9.7. Top row: Four patches of 35×35 pixels from the real data of Figure 9.6. Middle row: On the left we show three snapshots of the evolution of r_1 (the radiance of the background object), and on the right we show the true radiance captured after removing the occluding object from the scene. Bottom row: On the left we show three snapshots of the evolution of r_2 together with its support ϕ, and on the right we show the radiance captured as ground truth.

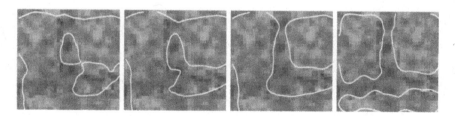

Figure 9.8. Real scene. Evolution of the support of the occluding object ϕ defined in equation (2.24). On the leftmost image we show the initial contour, and on the rightmost image we show the contour at convergence. The zero level set of the support function is shown superimposed on the pinhole image of the scene.

9.5 Summary

Modeling the shape of the scene with a single smooth surface is not appropriate in the presence of occluding boundaries. In fact, a real aperture lens can generate images where occluding portions of the scene "diffuse away" to reveal the

Figure 9.9. Real data. Four snapshots of the evolution of surfaces s_1 and s_2 together with the support ϕ defined in equation (2.24). The surface s_2 is shown only on the domain estimated via ϕ.

background that would not be visible with a pinhole model. In order to infer the radiance beyond occlusions it is necessary to model the occlusion process explicitly. This is easy to do conceptually by enriching the model with two radiances (occluder and occluded), two shapes, and a function that determines the spatial location of the occluder. Inference of the geometry and photometry of the scene can be done in the same conceptual manner as in previous chapters, by an alternating minimization, albeit a significantly more complex one, where guarantees of convergence to the ground truth scene are very difficult to provide. In practice, initialization with a single-depth model, and local iterations around the occluding boundaries, detected as the regions yielding high residual in the single-depth model, give empirically satisfying results, including the reconstruction of radiance behind occlusions.

10
Final remarks

Images are only an infinitesimal snapshot, in space and time, of the physical world. And yet they are a rich source of information that humans and other animals exploit to interact with it. The sophisticated variable-geometry of the lens in the human eye is known to play a role in the inference of spatial ordering, proximity and other three-dimensional cues. In engineering systems, accommodation artifacts such as defocus and motion blur are often seen as nuisances, and their effects minimized by means of expensive optics or image capture hardware. We hope that this book will at least encourage the engineer to look at defocus and motion blurs as friends, not foes.

Alone, defocus and motion blur provide three-dimensional information only under restricted circumstances, but depending on the optical characteristics of the imaging device, new devices can be created to better exploit these cues. For instance, most webcams have very small aperture lenses akin to pinhole cameras that do not allow resolving depth from defocus; however, an entire array of cameras can act as a large-aperture lens and resolve depth to a very fine degree, and overcome occlusions [Levoy et al., 2004]. Similarly, dedicated devices can be built to exploit this cue in the small scale, as is done in confocal microscopy [Ancin et al., 1996]. Regardless of the device, the combination of multiple low-quality images that are either defocused or motion blurred can be exploited to retrieve a super-resolution image, in both space and time, of higher quality than any of the original images.

We have devoted an entire chapter to the handling of occlusions. Occlusions, or more generally speaking visibility artifacts, play a very important role in the formation of images, and consequently also in their interpretation. They make image formation an intrinsically non-linear process, whereas in their absence a linear model would be sufficient to describe the energy transport phenomena involved. Occlusions ofter result in sharp discontinuity on the image. This is the basis for image segmentation as an analysis tool and the primary cause of the highly kurtotic nature of natural image statistics. However, image discontinuities can be due to occlusions, that have a direct spatial correlate, or simply to reflectance or illumination. We hope to have convinced the reader that accommodation cues provide significant help to disambiguate these two scenarios: occluding boundaries behave quite differently from albedo discontinuities in response to changes in accommodation. This should be exploited to the best possible extent.

Eventually, a successful and flexible artificial vision system will have to make use of and integrate all available cues. This book, therefore, is not meant to be used as a sole source on the retrieval of three-dimensional information from image sequences. Nevertheless, the analytical framework proposed is rather flexible and we expect that other cues can be integrated into it in the near future. We also expect that, as the speed of processing hardware advances, the algorithms described will be implemented in real-time and be used on the field to complement other cues and enhance their performance despite low-quality optical hardware.

We conclude with a note on curriculum. While there seems to be a trend in the modern US academia away from teaching mathematics in depth, especially as it concerns continuous mathematics in Computer Science departments, nevertheless as computers become "smarter" and interact more and more with the physical world, mathematical models of physical processes will once again be a relevant part of the curriculum. We have tried to maintain the use of mathematical notation at the lowest possible level, but some students may find the content challenging. However, we hope that it will be rewarding in the long term and that some of the tools learned in this book will prove useful in other endeavors as well.

Appendix A
Concepts of radiometry

The purpose of this appendix is to give the reader an overview of the basic notions of radiometry. In particular, we give the definitions of radiance, irradiance and bidirectional reflectance distribution function. Then, by using these notions, we will derive a simple imaging model for shape from defocus in the case of a thin lens (see Section A.2).

A.1   Radiance, irradiance, and the pinhole model

Let s be a surface in space. We denote the tangent plane to the surface at a point p by $dA(p)$ and its outward unit normal vector by $\nu(p)$ (see Figure A.1). Let L be a surface that is irradiating light, which we call the *light source*. For simplicity, we may assume that L is the only source of light in space. At a point $q \in L$, we denote $dA(q)$ and $\nu(q)$ the tangent plane and the outward unit normal of L, respectively, as shown in Figure A.1.

A.1.1   Foreshortening and solid angle

Before defining how light and surfaces interact, we need to introduce the notion of *foreshortening* and that of *solid angle*. Foreshortening gives a measure of the effective light energy hitting a surface, depending on the surface orientation with respect to the source of illumination. In formulas, if $dA(p)$ is the infinitesimal surface area at p, and $l(p)$ is the unit vector that indicates the direction from p to

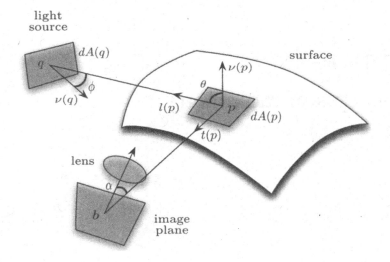

Figure A.1. Geometry of the scene. A light source emits energy towards a surface, which reflects part of it towards the image plane.

q (see Figure A.1), then the corresponding foreshortened area as seen from q is

$$\cos(\theta)dA(p), \tag{A.1}$$

where θ is the angle between the direction $l(p)$ and the normal vector $\nu(p)$. The solid angle formed by a point q and a surface $dA(p)$, is defined to be the projection of $dA(p)$ through the point q on the unit sphere centered at the point q. Then, the infinitesimal solid angle $d\omega(q)$ made by the point q and the infinitesimal area $dA(p)$ is

$$d\omega(q) \doteq \frac{\cos(\theta)}{\|p-q\|^2}dA(p), \tag{A.2}$$

where $\|p-q\|$ is the distance between p and q.

A.1.2   Radiance and irradiance

The *radiance* is defined to be the amount of energy emitted along a certain direction, per unit area perpendicular to the direction of emission (the foreshortening effect), per unit solid angle, and per unit time, following the definition in [Forsyth and Ponce, 2003]. According to our notation, if we denote the radiance at the point q in the direction of p by $r(q, l(p))$, the energy dE emitted by the light L at a point q towards p on s is

$$dE(q, l(p)) \doteq r(q, l(p)) \cos(\phi) \, dA(q) \, d\omega(q) \, dt, \tag{A.3}$$

where $\cos(\phi)dA(q)$ is the foreshortened area of $dA(q)$ seen from the direction of p, and $d\omega(q)$ is the solid angle given in equation (A.2).

Although the radiance is used for energy that is emitted, a more appropriate measure for incoming energy is called *irradiance*. The irradiance is defined as the amount of energy received along a certain direction, per unit area, and per unit time. Notice that in the case of the irradiance, we do not foreshorten the surface area as in the case of the radiance. Denote the irradiance at p in the direction $l(p)$ by $I(p, l(p))$. By energy preservation, we have $I(p, l(p))dA(p)dt = dE(q, l(p))$. Then the radiance r at a point q that illuminates the surface $dA(p)$ and the irradiance I measured at the same surface $dA(p)$ are related by

$$I(p, l(p)) = r(q, l(p)) \cos(\theta) \, d\omega(p), \tag{A.4}$$

where $d\omega(p) = dA(q) \cos(\phi)/\|p - q\|^2$ is the solid angle of $dA(q)$ seen from p.

A.1.3   Bidirectional reflectance distribution function

The portion of energy coming from a direction $l(p)$ that is reflected onto a direction $t(p)$ by the surface s (see Figure A.1) is described by $\rho(t(p), l(p))$, the *bidirectional reflectance distribution function* (BRDF). Here both $t(p)$ and $l(p)$ are normal vectors. More precisely, if $dr(p, t(p), l(p))$ is the amount of radiance emitted in the direction $t(p)$ due to the irradiance $I(p, l(p))$, then, the BRDF is given by the ratio

$$\rho(t(p), l(p)) \doteq \frac{dr(p, t(p), l(p))}{I(p, l(p))} = \frac{dr(p, t(p), l(p))}{r(q, l(p)) \cos(\theta) \, d\omega(p)}. \tag{A.5}$$

To obtain the total radiance at a point p in the outgoing direction $t(p)$, we need to integrate the BRDF against all the incoming irradiance directions $l(p)$ in the hemisphere Γ at p:

$$r(p, t(p)) = \int_\Gamma dr(p, t(p), l(p)) = \int_\Gamma \rho(t(p), l(p)) \, r(q, l(p)) \cos(\theta) \, d\omega(p). \tag{A.6}$$

A.1.4   Lambertian surfaces

The above model can be considerably simplified if we restrict our attention to a class of materials, called *Lambertian*, that do not change appearance depending on the viewing direction. For a perfect Lambertian surface, its radiance $r(p, t(p))$ only depends on how the surface faces the light source, but not on the direction $t(p)$ from which it is viewed. Therefore, $\rho(t(p), l(p))$ is actually independent of $t(p)$ (and by the Helmholtz reciprocity principle, also independent of $l(p)$ [Forsyth and Ponce, 2003]). More precisely, for Lambertian surfaces, we have

$$\rho(t(p), l(p)) = \rho(p), \tag{A.7}$$

where $\rho(p) : \mathbb{R}^3 \mapsto [0, \infty)$ is a scalar function. Hence, the radiance from the point p on a Lambertian surface s is

$$r(p) = \int_\Gamma \rho(p) \, r(q, l(p)) \cos(\theta) \, d\omega(p). \tag{A.8}$$

Notice that if the scene (surface and light) is static, and we do not change the viewing position, we can write the surface s explicitly as a function that assigns a depth value at each point $y = [y_1\ y_2]^T \in \Omega$ on the image plane; that is, $s : y \mapsto s(y)$. Then, we can also write the coordinates of a point p on the surface s in parametric form with respect to the image plane as

$$p = \begin{bmatrix} y \\ s(y) \end{bmatrix} \in \mathbb{R}^3. \tag{A.9}$$

Therefore, the radiance can also be represented by a function $r : y \mapsto r(y)$. We use this notation throughout the book and, in particular, in Section A.2.

A.1.5 Image intensity for a Lambertian surface and a pinhole lens model

Suppose that our imaging sensor is well modeled by a pinhole lens. Then, by measuring the amount of energy received along the direction $t(p)$, the irradiance I at b can be expressed as a function of the radiance from the point p:

$$I(b, t(p)) = r(p) \frac{\pi}{4} \left(\frac{D}{v} \right)^2 \cos^4(\alpha), \tag{A.10}$$

where D is the lens diameter, v is the distance between the lens and the image plane (i.e., the focus setting), and α is the angle between the optical axis (i.e., the normal to the image plane) and the direction $t(p)$. A detailed derivation of the above formula can be found in [Horn, 1987].

A.2 Derivation of the imaging model for a thin lens

In this section, we are interested in deriving an expression for the irradiance measured on the image plane when the optics are modeled as a *thin lens*. Consider a point $\bar{y}$ on the image plane, and a finite surface δA_0, corresponding to the area of a CCD element (see Figure A.2). We denote with r the radiance defined on the surface s. Because the radiance emitted by a point $\tilde{x}$ towards a point x is the same as the radiance emitted by a point x towards a point $\tilde{x}$, the radiance at a point x on the lens is $r(\tilde{x})$. The radiance $r(\tilde{x})$ that each point x on the lens irradiates towards the point $\bar{y}$ has to be integrated over the solid angle $\delta w(\bar{y})$ defined by the area δA_0 on $\bar{y}$ and the point x, and over the foreshortened infinitesimal area $\cos(\psi) dA_1$ around x. This energy divided by the surface element δA_0 gives the irradiance measured at the infinitesimal area dA_1 around x:

$$dI(\bar{y}) = \frac{r(\tilde{x}) \delta w(\bar{y}) \cos(\psi) dA_1}{\delta A_0} = r(\tilde{x}) \frac{\cos^2(\psi)}{\|x - \bar{y}\|^2} dA_1 = r(\tilde{x}) \frac{\cos^4(\psi)}{v^2} dA_1, \tag{A.11}$$

where ψ is the angle between the normal to the image plane and the vector from $\bar{y}$ to x, and v is the focus setting, that is, the distance between the image plane and

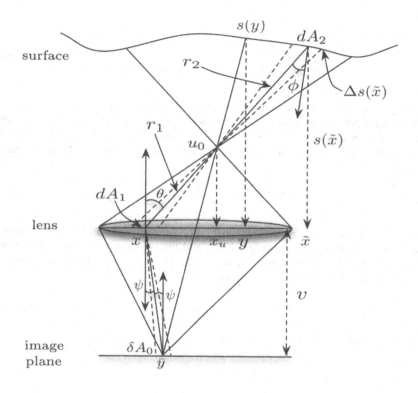

Figure A.2. Simplified geometry of a thin lens imaging a scene made of a single surface.

the lens plane. Then, the irradiance measured at a point $\bar{y}$ is given by the integral of all the single contributions from each point on the lens. The integral can be written as

$$I(\bar{y}) = \int_{\Psi} r(\tilde{x}) \frac{\cos^4(\psi)}{v^2} dA_1, \qquad (A.12)$$

where Ψ is the domain defined by the lens (of diameter D). If we consider the radiance r to be a constant, and the angle ψ independent of the coordinates x on the image plane, then equation (A.12) becomes

$$I(\bar{y}) = r \frac{\pi}{4} \left(\frac{D}{v} \right)^2 \cos^4(\psi) \qquad (A.13)$$

which resembles equation (A.10) obtained in the case of a pinhole camera model. Notice that in the above equation irradiance and radiance are related by a scale factor. In our derivation, we redefine the radiance such that the scale factor is 1, and we use the approximation $\cos^4(\psi) = 1$; that is, instead of equation (A.12) we use

$$I(\bar{y}) = \frac{\int_\Psi r(\tilde{x}) dA_1}{\int_\Psi dA_1} = \frac{4}{\pi D^2} \int_\Psi r(\tilde{x}) dA_1. \tag{A.14}$$

To obtain an explicit expression for equation (A.14) we establish the relationship between the infinitesimal area on the lens plane dA_1 and the infinitesimal area on the surface dA_2. We start from the equivalence between the solid angles of the infinitesimal area dA_1 and the infinitesimal area dA_2 as seen from the point $[x_u \; u_0]^T$:

$$\frac{dA_1 \cos(\theta)}{r_1^2} = \frac{dA_2 \cos(\phi)}{r_2^2}, \tag{A.15}$$

where the angles θ and ϕ are shown in Figure A.2, the segment r_1 is the distance between the points $[x \; 0]^T$ and $[x_u \; u_0]^T$, and the segment r_2 is the distance between the points $[\tilde{x} \; s(\tilde{x})]^T$ and $[x_u \; u_0]^T$. The point x_u satisfies $x_u = -\bar{y}u_0/v$, and the point x can be written in terms of $\tilde{x}$ as

$$x = x_u + u_0 \frac{x_u - \tilde{x}}{s(\tilde{x}) - u_0}. \tag{A.16}$$

Define the vector

$$V = \begin{bmatrix} x - x_u \\ -u_0 \end{bmatrix} \tag{A.17}$$

and the vector normal to the surface s by:

$$N = \begin{bmatrix} \nabla s(\tilde{x}) \\ -1 \end{bmatrix}, \tag{A.18}$$

where ∇s is the two-dimensional gradient of the surface s with respect to the lens plane. From $r_1^2 = \|V\|^2$, we have the following equivalences.

$$\cos(\theta) = \frac{u_0}{\|V\|}$$
$$\cos(\phi) = \frac{|\langle N, V\rangle|}{\|N\|\|V\|} = \frac{|u_0 + \nabla s(\tilde{x})^T(x - x_u)|}{\|V\|\sqrt{1 + \|\nabla s(\tilde{x})\|^2}} \tag{A.19}$$

and we are left with computing the segment r_2:

$$r_2^2 = \left\| \begin{matrix} \tilde{x} - x_u \\ s(\tilde{x}) - u_0 \end{matrix} \right\|^2 = \|V\|^2 \frac{(s(\tilde{x}) - u_0)^2}{u_0^2}. \tag{A.20}$$

Substituting, we have:

$$dA_1 = \frac{|s(\tilde{x}) - u_0 + \nabla s(\tilde{x})^T(x_u - \tilde{x})|u_0^2}{\sqrt{1 + \|\nabla s(\tilde{x})\|^2}|s(\tilde{x}) - u_0|^3} dA_2 \tag{A.21}$$

or, by introducing the expression of x_u in terms of $\bar{y}$:

$$dA_1 = \frac{|s(\tilde{x}) - u_0 - \nabla s(\tilde{x})^T(\frac{\bar{y}u_0}{v} + \tilde{x})|u_0^2}{\sqrt{1 + \|\nabla s(\tilde{x})\|^2}|s(\tilde{x}) - u_0|^3} dA_2. \tag{A.22}$$

Now, because $dA_2 = \|N\|d\tilde{x}$, we have:

$$dA_1 = \frac{|s(\tilde{x}) - u_0 - \nabla s(\tilde{x})^T(\frac{\bar{y}u_0}{v} + \tilde{x})|u_0^2}{|s(\tilde{x}) - u_0|^3} d\tilde{x}. \tag{A.23}$$

Then, the original imaging model becomes:

$$I(\bar{y}) = \int_{\Psi} r(\tilde{x}) \frac{4u_0^2}{\pi D^2} \frac{|s(\tilde{x}) - u_0 - \nabla s(\tilde{x})^T(\frac{\bar{y}u_0}{v} + \tilde{x})|}{|s(\tilde{x}) - u_0|^3} d\tilde{x}, \tag{A.24}$$

where the domain $\Psi = \{x : \|x\| \leq D/2\}$ can be written as

$$\left\| \bar{y}\frac{s(\tilde{x})}{v} + \tilde{x} \right\|^2 \leq \frac{D^2}{4} \frac{(s(\tilde{x}) - u_0)^2}{u_0^2}. \tag{A.25}$$

As observed in Chapter 2, Section 2.2 if we neglect the linear term $\nabla s(\tilde{x})$ we obtain the classical equifocal model with a pillbox kernel.

Appendix B
Basic primer on functional optimization

Several times throughout the book we have had to face the problem of finding the minimizer of a cost functional, typically called E to remind the reader of the physical analogy with energy. The variables with respect to which the energy is to be minimized are typically the unknown shape s, as in equation (6.25), sometimes the unknown radiance r, as in equation (5.2), and yet other times the velocity field ν. For the sake of example, let us restrict our attention to s only; the reader can easily extend the arguments to the other variables.

If these unknowns lived in a finite-dimensional vector space, say $\mathbb{R}^n$, then even for the case of non-linear energy functions $E(s)$, one could easily devise an iterative minimization procedure that, starting from an initial condition s_0, would converge at least to a local minimum of the function E. We recall that a local minimum $\hat{s}$ is a point such that, in a neighborhood O of $\hat{s}$, we have $E(\hat{s}) \leq E(s) \ \forall s \in O$. Under suitable regularity conditions of the function E and open-ness of the domain O, a minimum is attained where the gradient of E vanishes; that is, $\nabla E(\hat{s}) = 0$. In other words, $\nabla E(\hat{s}) = 0$ is a necessary, but not sufficient, condition for $\hat{s}$ to be a minimum. This, however, can be used to design an iteration that has $\hat{s}$ as its fixed point, provided that we start within the neighborhood O of the minimum, for instance,

$$s_{k+1} = s_k - \beta_k \nabla E(s_k) \qquad k \geq 0, \tag{B.1}$$

where $s_0 \in O$ and $\beta_k > 0$ could be a constant, or a function of k, designed so that the sequence $E(s_k)$ diminishes at each iteration; that is, $E(s_{k+1}) \leq E(s_k)$. The reader who is not familiar with this type of optimization in the finite-dimensional case can consult one of many textbooks, for instance [Polak, 1997], Chapter 1, where the algorithm thus described is called Algorighm 1.3.1 and summarized on page 56.

Unfortunately, our unknowns do not live in a finite-dimensional vector space, like $\mathbb{R}^n$, but are instead *functions*, $s : \Omega \subset \mathbb{R}^2 \rightarrow \mathbb{R}$. The extension of the simple descent algorithm above to the case of functional unknowns is the subject of the *calculus of variations*. Our goal in this appendix is not to give a rigorous and comprehensive introduction to the calculus of variations, for which the reader can consult a number of books, such as [Fomin and Gelfand, 2000]. Instead, we intend to present the bare minimum that allows the reader to fully appreciate the calculations done to derive the descent algorithms used throughout the book. In the end, our algorithms look a lot like (B.1), except that we need to be careful to understand the meaning of the gradient of a function(al) E with respect to a function s. We do this in the next section. In the following sections, we will report the detailed calculations of the gradients of the cost functionals employed in the various chapters of the book. We encourage the reader to go through the exercise of deriving these gradients at least once, so that the calculations detailed here can serve for verification.

B.1 Basics of the calculus of variations

This section is meant to give the reader an intuitive understanding of the basic ideas of the calculus of variations as they pertain to the derivation of the algorithms reported throughout the book. As we only present the details necessary to derive the gradients used in our algorithms, we encourage the reader who is interested in a deeper understanding of these concepts to consult [Fomin and Gelfand, 2000] or an equivalent textbook.

The general goal in this section is to minimize a real-valued function E of another function s with respect to the latter. Such functions of functions are also called *functionals*. In particular, we are interested in functionals of the form

$$E(s) = \int_\Omega \varphi(s, \nabla s, y) dy, \tag{B.2}$$

where $\Omega \subset \mathbb{R}^2$, $\varphi \geq 0$, and $s : \Omega \rightarrow \mathbb{R}$ belongs to a normed space S where a suitable norm $\|s\|$ has been defined. Unlike the case of $\mathbb{R}^n$, there is no obvious choice of norm, and indeed different norms are appropriate for different applications, but we do not enter this discussion here and refer the reader to the literature for details. All we need at this stage is to know that this norm allows us to define a distance between two points on S, say s_1 and s_2, via $\|s_2 - s_1\|$. We do, however, assume for the sake of simplicity that s is a smooth function, so that its derivatives (of high enough order) are always defined. We assume that the space S is linear; that is, such that one can add any two elements in the space, say $s \in S$ and $h \in S$, to obtain another element $s + h \in S$ in the same space.

We now introduce the notion of the (Fréchet, or functional) derivative of the functional E.

B.1.1   Functional derivative

Consider the following increment, obtained by perturbing the function s with another function h,

$$\Delta E = E(s + h) - E(s). \qquad \text{(B.3)}$$

If s is a fixed function, the increment ΔE can be thought of as a functional of the perturbation h, $\Delta E(h)$, and in general it is a nonlinear one. If one can isolate the linear part of this functional, which we call δE, in such a way that

$$\Delta E(h) = \delta E(h) + \epsilon \|h\| \qquad \text{(B.4)}$$

and $\epsilon \to 0$ as $\|h\| \to 0$, then the functional E is called *differentiable* and δE is its differential, or *first-order variation*. One can think of defining the differential as the limit

$$\delta E \doteq \lim_{\|h\| \to 0} \frac{E(s + h) - E(s)}{\|h\|} \qquad \text{(B.5)}$$

when this limit exists. Note that δE is a function of both s and h. In fact, it is by construction a linear function of h, so we can write it as

$$\delta E(s, h) = \int_\Omega \nabla_s E(s)(y) h(y) dy, \qquad \text{(B.6)}$$

where $\nabla_s E(s)$ is a function called the Fréchet (or functional) derivative. Because it generalizes the notion of gradient ∇E familiar to the reader when s is a finite-dimensional vector, we call $\nabla_s E(s)$ the gradient of the functional E.

We are now ready to extend the notion of iterative descent optimization to functionals. To that end, we recall that a local minimum $\hat{s}$ for a functional E is a function such that $E(\hat{s}) \leq E(s)$ for all s in a neighborhood [1] $\{s \mid \|s - \hat{s}\| \leq \epsilon\}$. Now, it can be shown (see, e.g., [Luenberger, 1968], Theorem 1, page 178), that a necessary condition for a functional E to have a local minimum at $\hat{s}$ is that the differential $\delta E(s, h) = 0$ vanishes for all admissible variations h. Therefore, a necessary condition is that the functional derivative $\nabla_s E(\hat{s})$ also vanishes at the minimum. This is the basis for the iterative algorithms described in this book, where the iteration now involves functions, rather than points on a finite-dimensional vector space

$$s_{k+1} = s_k - \beta_k \nabla_s E(s_k). \qquad \text{(B.7)}$$

We now write the functional derivative specifically for functionals of the form (B.2), to arrive at Euler–Lagrange equations.

---

[1] In practice one has to be careful that the variation $h \doteq s - \hat{s}$ is admissible; that is, that it satisfies the constraints imposed on the space S. However, in this appendix we only wish to give the reader an intuitive understanding of the calculus of variation and therefore we glance over these more technical issues.

B.1.2   Euler–Lagrange equations

In this section we are interested in computing the variation of the functional E in (B.2). We do the derivation for a scalar function $s : \Omega \subset \mathbb{R} \to \mathbb{R}$, for simplicity, and let the reader extend it to several variables. So, the integral in (B.2) is on an interval of the real line, and the variations h that we consider have to vanish at the boundaries of this interval unless we want to consider the interval itself to be an unknown. As we have indicated in the previous section, the necessary condition for $\hat{s}$ to be a minimum is that the differential $\delta E(s, h)$ vanishes for all admissible variations h. Given that the functional E is Fréchet differentiable, we can compute the differential $\delta E(s, h)$ as [2]

$$\delta E(s, h) = \frac{d}{d\epsilon} \int_\Omega \varphi(s + \epsilon h, \dot{s} + \epsilon \dot{h}, y) dy \Big|_{\epsilon=0} \qquad (\text{B.8})$$

for an arbitrary $h \in S$ such that $h(x) = 0, \forall x \in \partial\Omega$, and a scalar ϵ. Because we are dealing with scalar functions, ordinary derivatives are indicated by a dot. This can be broken down into two terms, using the definition of the functional derivative:

$$\delta E(s, h) = \int_\Omega \nabla_s \varphi(s, \dot{s}, y) h(y) dy + \int_\Omega \nabla_{\dot{s}} \varphi(s, \dot{s}, y) \dot{h}(y) dy. \qquad (\text{B.9})$$

Under suitable regularity assumptions, we can integrate this expression by parts to obtain

$$\delta E(s, h) = \int_\Omega \left(\nabla_s \varphi(s, \dot{s}, y) - \frac{d}{dy} \nabla_{\dot{s}} \varphi(s, \dot{s}, y) \right) h(y) dy + \nabla_{\dot{s}} \varphi(s, \dot{s}, y) h(y) \Big|_{\partial\Omega}. \qquad (\text{B.10})$$

Because the variation h is zero at the boundary of the interval Ω, the second term vanishes and one can identify the functional derivative $\nabla_s E$ as

$$\nabla_s E(y) = \nabla_s \varphi(s, \dot{s}, y) - \frac{d}{dy} \nabla_{\dot{s}} \varphi(s, \dot{s}, y). \qquad (\text{B.11})$$

Again under a suitable regularity assumption, because $\delta E(s, h)$ has to be zero for any admissible variation h, one can conclude that the functional derivative $\nabla_s E$ has to be zero. This is the Euler–Lagrange equation:

$$\nabla_s \varphi(s, \dot{s}, y) - \frac{d}{dy} \nabla_{\dot{s}} \varphi(s, \dot{s}, y) = 0. \qquad (\text{B.12})$$

Again, we have glanced over the conditions under which this can be established rigorously. These are established in three lemmas on pages 9 and 10 of [Fomin and Gelfand, 2000]. The derivation of the multivariable case is also derived in Section 5 of the same reference.

---

[2] The differential given in equation (B.8) is more formally introduced as the *Gâteaux* differential [Fomin and Gelfand, 2000].

B.2   Detailed computation of the gradients

In this section we report the detailed calculations of the functional derivatives necessary to implement the algorithms described throughout this book.

B.2.1   Computation of the gradients in Chapter 6

To compute the gradients (6.29) we need to derive the first-order variation of the cost functional (6.25). In the case of

$$E_3 = \int_\Omega \|\nabla s(y)\|^2 dy \tag{B.13}$$

the computation amounts to:

$$\delta E_3(s, h) = \int_\Omega -2(\nabla \cdot \nabla s(y))h(y)dy = 0 \tag{B.14}$$

for an arbitrary h such that $h(y) = 0, \forall y \in \partial\Omega$. Hence, the functional derivative with respect to E_3 is

$$\nabla E_3(y) = -2\triangle s(y). \tag{B.15}$$

The other two derivatives relative to E_1 and E_2 are similar to each other, so we derive only the one in E_1. The first complication we meet in computing the functional derivative relative to E_1 is due to the fact that the function u is not given in explicit form, but as the solution of a PDE. When $y \in \Omega_+$, u is the solution of

$$\begin{cases} \dot{u}(y, t) = \nabla \cdot (|c(y)|\nabla u(y, t)) & t \in [0, \infty) \\ u(y, 0) = J_1(y) & \forall y \in \Omega_+ \\ \\ u(y, \Delta t) = J_2(y) & \forall y \in \Omega_+ \end{cases} \tag{B.16}$$

The first-order variation [Luenberger, 1968] of the term

$$E_1 \doteq \int_{\Omega_+} (u(y, \Delta t) - J_2(y))^2 dy \tag{B.17}$$

is

$$\delta E_1(s, h) = 2\int_{\Omega_+} (u(y, \Delta t) - J_2(y))\delta u(y, \Delta t)dy + \delta(c(y))(u(y, \Delta t) - J_2(y))^2 \tag{B.18}$$

for an arbitrary function $h : \Omega_+ \to \mathbb{R}$ such that $h(y) = 0, \forall y \in \partial\Omega_+$. Now, defining

$$\delta u(y, t) \doteq \lim_{\epsilon \to 0} \frac{u(y, t)\big|_{c+\epsilon h} - u(y, t)\big|_c}{\epsilon} \tag{B.19}$$

we have that δu satisfies

$$\begin{cases} \delta\dot{u}(y, t) = \nabla \cdot (c(y)\nabla\delta u(y, t) + h(y)\nabla u(y, t)) & t \in (0, \infty) \\ \delta u(y, 0) = 0 & \forall y \in \Omega_+ \end{cases} \tag{B.20}$$

where the diffusion coefficient c satisfies equation (6.17). Let $z : \Omega \times [0, \infty) \to \mathbb{R}$ be a function such that

$$z(y, \Delta t) = u(y, \Delta t) - J_2(y).$$
(B.21)

Then, by substituting z in equation (B.18), we obtain:

$$
\begin{aligned}
\delta E_1(s, h) \ = \ & 2 \int_{\Omega_+} z(y, \Delta t) \delta u(y, \Delta t) dy \\
& + \delta(c(y))(u(y, \Delta t) - J_2(y))^2 \\
= \ & 2 \int_{\Omega_+} z(y, 0) \delta u(y, 0) dy + \delta(c(y))(u(y, \Delta t) - J_2(y))^2 \\
& + 2 \int_0^{\Delta t} \int_{\Omega_+} \dot{z}(y, t) \delta u(y, t) + z(y, t) \delta \dot{u}(y, t) dy dt \\
& + \delta(c(y))(u(y, \Delta t) - J_2(y))^2
\end{aligned}
$$
(B.22)

which becomes

$$
\begin{aligned}
\delta E_1(s, h) \ = \ & 2 \int_0^{\Delta t} \int_{\Omega_+} \Big(\dot{z}(y, t) + \nabla \cdot (c(y) \nabla z(y, t)) \Big) \delta u(y, t) \\
& - h(y) \nabla z(y, t) \cdot \nabla u(y, t) dy dt \\
& + \delta(c(y))(u(y, \Delta t) - J_2(y))^2
\end{aligned}
$$
(B.23)

after integrating by parts twice and using the fact that both h and the diffusion coefficient c vanish at the boundary of Ω_+, and noticing that the initial conditions of δu are $\delta u(y, 0) = 0$. To simplify the above equation we choose the function z such that it satisfies

$$
\begin{cases}
\dot{z}(y, t) = -\nabla \cdot (c(y) \nabla z(y, t)) & t \in (0, \Delta t] \\
z(y, \Delta t) = u(y, \Delta t) - J_2(y) & \forall y \in \Omega_+.
\end{cases}
$$
(B.24)

The substitution of z in equation (B.23) yields

$$
\begin{aligned}
\delta E_1(s, h) \ = \ & -2 \int_0^{\Delta t} \int_{\Omega_+} h(y) \nabla z(y, t) \cdot \nabla u(y, t) dy dt \\
& + \delta(c(y))(u(y, \Delta t) - J_2(y))^2.
\end{aligned}
$$
(B.25)

Now, define $w(y, t) = z(y, \Delta t - t)$, the *adjoint equation* (also notice that this is the definition of equations (6.30) and (6.33)). Then, we immediately obtain

$$
\begin{cases}
\dot{w}(y, t) = \nabla \cdot (c(y) \nabla w(y, t)) & t \in (0, \Delta t] \\
w(y, 0) = u(y, \Delta t) - J_2(y) & \forall y \in \Omega_+
\end{cases},
$$
(B.26)

which, substituted in the expression of δE_1, gives

$$
\begin{aligned}
\delta E_1(s, h) \ = \ & -2 \int_0^{\Delta t} \int_{\Omega_+} h(y) \nabla w(y \Delta t - t) \cdot \nabla u(y, t) dy dt \\
& + \delta(c(y))(u(y, \Delta t) - J_2(y))^2.
\end{aligned}
$$
(B.27)

Now, the gradient ∇E_1 is easily determined as

$$
\begin{aligned}
\nabla E_1(y) \;=\; & -2H(c(y)) \int_0^{\Delta t} \nabla w(y, \Delta t - t) \cdot \nabla u(y,t)dt \\
& +\delta(c(y))(u(y, \Delta t) - J_2(y))^2.
\end{aligned}
\tag{B.28}
$$

Notice that both functions u and w can be computed by simulating the respective models, and both models involve only forward diffusions.

B.2.2   Computation of the gradients in Chapter 7

We formally compute directional derivatives in direction (h_r, h_s, h_V) in the following. The first-order variation of the objective functional (7.13) is given by

$$
\begin{aligned}
\delta E(r, s, V_{1,2}) \;=\; & 2 \sum_{i=1}^{K} \int_\Omega (u_i(x, t_1) - J(x, v_i))\delta u_i(x, t_1)dx \\
& + 2\alpha_0 \int_\Omega (r(x) - r^*(x))h_r(x)dx \\
& + 2\alpha_1 \int_\Omega \nabla s(x) \cdot \nabla h_s(x)dx \\
& + 2\alpha_2(|V_{1,2}| - M_0)\frac{V_{1,2} \cdot h_V}{|V_{1,2}|},
\end{aligned}
\tag{B.29}
$$

where, similarly to the previous section, δu_i is the solution of the initial value problem

$$
\begin{cases}
\delta \dot{u}_i(x, t) = \nabla \cdot (D_i(x)\nabla \delta u_i(x,t) + \delta D(x)\nabla u_i(x,t)) & \\
\delta u_i(x, 0) = h_r(x) & t \in (0, \infty) \;, \\
(D_i(x)\nabla \delta u_i(x,t) + \delta D(x)\nabla u_i(x,t)) \cdot n(x) = 0 & \forall x \in \partial\Omega
\end{cases}
\tag{B.30}
$$

the diffusion tensor D_i is given by equation (7.7), and δD_i is the variation of the diffusion tensor, given by

$$
\delta D_i(x) = B_i(x)h_s(x) + C_i(x)h_V,
\tag{B.31}
$$

where

$$
\begin{aligned}
B_i(x) \;=\; & \sigma_i(x)\frac{\gamma D v_i}{s^2(x)}I_d - 2\frac{v_i^2 \Delta \tilde{T}^2}{s^3(x)}\left(V_{1,2}V_{1,2}^T\right) \\
C_i(x) \;=\; & \frac{v_i^2 \Delta \tilde{T}^2}{s^2(x)}\left(\begin{bmatrix} 1 \\ 1 \end{bmatrix} V_{1,2}^T + V_{1,2}[1\ 1]\right).
\end{aligned}
\tag{B.32}
$$

The computation of the functional derivative relative to the term δE_2,

$$
\delta E_2(x) \doteq 2\alpha_0 \int_\Omega (r(x) - r^*(x))h_r(x)dx
\tag{B.33}
$$

yields only one term different from 0,

$$
\nabla_r E_2(x) = 2\alpha_0(r(x) - r^*(x)).
\tag{B.34}
$$

Similarly, the computation of functional derivative relative to the term δE_3,

$$\delta E_3(x) \doteq 2\alpha_1 \int_\Omega \nabla s(x) \cdot \nabla h_s(x) dx \qquad (B.35)$$

yields only one term different from 0,

$$\nabla_s E_3(x) = -2\alpha_1 \triangle s(x) \qquad (B.36)$$

and again the computation of the functional derivative relative to the term δE_4,

$$\delta E_4(x) \doteq 2\alpha_2(|V_{1,2}| - M_0) \frac{V_{1,2} \cdot h_V}{|V_{1,2}|} \qquad (B.37)$$

yields only one term different from 0,

$$\nabla_V E_4(x) = 2\alpha_2(|V_{1,2}| - M_0) \frac{V_{1,2}}{|V_{1,2}|}. \qquad (B.38)$$

In the following, we report the more involved computation of the term δE_1,

$$\delta E_1 \doteq 2 \sum_{i=1}^{K} \int_\Omega (u_i(x,t_1) - J(x,v_i))\delta u_i(x,t_1) \, dx. \qquad (B.39)$$

In the previous section, in order to avoid the expensive computation of the gradient by solving a time-dependent problem for each variation, we employed the adjoint method, whose main idea is to simplify the gradient computation by introducing an adjoint system of differential equations. We do the same also in this case, and the adjoint problem is given by

$$\begin{cases} \dot{w}_i(x,t) = \nabla \cdot (D_i(x)\nabla w_i(x,t)) \\ w_i(x,0) = u_i(x,t_1) - J(x,v_i) \\ (D_i(x)\nabla w_i(x,t)) \cdot n(x) = 0. \end{cases} \qquad (B.40)$$

Using Gauss' theorem, we now obtain

$$\begin{aligned}
\delta E_1 &= 2 \sum_{i=1}^{K} \int_\Omega w_i(x,0)\delta u_i(x,t_1) \, dx \\
&= 2 \sum_{i=1}^{K} \int_\Omega w_i(x,t_1)h_r(x) \, dx \\
&\quad + 2 \sum_{i=1}^{K} \int_0^{t_1}\!\!\int_\Omega (\dot{w}_i(x,t_1-t)\delta u_i(x,t) \\
&\quad + w_i(x,t_1-t)\delta\dot{u}_i(x,t)) \, dx \, dt \\
&= 2 \sum_{i=1}^{K} \int_\Omega w_i(x,t_1)h_r(x) \, dx \\
&\quad - 2 \sum_{i=1}^{K} \int_0^{t_1}\!\!\int_\Omega (\delta D_i(x)\nabla u_i(x,t)) \cdot \nabla w_i(x,t_1-t) \, dx \, dt \\
&= \int_\Omega \nabla_r E_1(x)h_r(x) + \nabla_s E_1(x)h_s(x) \, dx + \nabla_{V_{1,2}} E_1^T h_V.
\end{aligned} \qquad (B.41)$$

Using the formula for $\delta D_i(x)$ we can compute the gradients of equation (7.13) as

$$
\begin{aligned}
\nabla_r E(x) &= 2 \sum_{i=1}^{K} w_i(x, t_1) + 2\alpha_0(r(x) - r^*(x)) \\
\nabla_s E(x) &= -2 \sum_{i=1}^{K} \int_0^{t_1} (B_1(x)\nabla u_i(x, t)) \cdot \nabla w_i(x, t_1 - t)\, dt \\
&\quad - 2\alpha_1 \triangle s(x) \\
\nabla_{V_{1,2}} E &= -2 \sum_{i=1}^{K} \int_0^{t_1} \int_\Omega (C_i(x)\nabla u_i(x, t)) \cdot \nabla w_i(x, t_1 - t)\, dx\, dt \\
&\quad + 2\alpha_2(|V_{1,2}| - M_0)\frac{V_{1,2}}{|V_{1,2}|}.
\end{aligned}
$$
(B.42)

The computation of the gradients involves the solution u_i of equation (7.9) with tensors D_i, and, subsequently, the solution w_i of equation (B.40). Both of these steps are well-posed due to Theorem 7.1. Thus, the gradients exist if the terms in equation (B.42) are finite.

B.2.3 Computation of the gradients in Chapter 8

Gradients of equation (8.16)

The computation of the gradients of the energy functional (8.16) follows closely the gradient computation of the previous section. However, to keep Chapter 8 as self contained as possible, we derive them again. We formally compute directional derivatives in direction $(h_s, h_{V^1}, h_{V^2}, h_\phi)$ in the following. The first-order variation of the objective functional (8.16) is given by

$$
\begin{aligned}
\delta E &= 2 \int H(\phi(x))(u_1(x, t_1) - J(x, \Delta \tilde{T}_2))\delta u_1(x, t_1)dx \\
&\quad + 2 \int H(-\phi(x))(u_2(x, t_1) - J(x, \Delta \tilde{T}_2))\delta u_2(x, t_1)dx \\
&\quad + 2\alpha_0 \int_\Omega \nabla s(x) \cdot \nabla h_s(x)dx \\
&\quad + 2\alpha_1 \int_\Omega \left(\int_\Omega s(y)dy - s^{mean} \right) h_s(x)dx \\
&\quad + 2\alpha_2 \int_{\partial\Omega_1} \frac{\nabla\phi}{\|\nabla\phi(x)\|} \cdot \nabla h_\phi(x)dx,
\end{aligned}
$$
(B.43)

where, similarly to the previous section, δu_j is the solution of the initial value problem

$$
\begin{cases}
\delta \dot{u}_j(x, t) = \nabla \cdot (D_j(x)\nabla\delta u_j(x, t) + \delta D(x)\nabla u_j(x, t)) \\
\delta u_j(x, 0) = 0 & t \in (0, \infty) \\
(D_j(x)\nabla\delta u_j(x, t) + \delta D_j(x)\nabla u_j(x, t)) \cdot n(x) = 0 & \forall x \in \partial\Omega
\end{cases}
$$
(B.44)

with diffusion tensor D_j given in equation (8.15) and δD_j is the variation of the diffusion tensor, given by,

$$\delta D_j(x) = B_j(x)h_s(x) + C_j(x)h_{V_j}. \tag{B.45}$$

where

$$
\begin{aligned}
B_j(x) &= -2\frac{v^2\left(\Delta\tilde{T}_2^2 - \Delta\tilde{T}_1^2\right)}{s^3(x)}V_{1,2}^j(V_{1,2}^j)^T \\
C_j(x) &= 2\frac{v^2\left(\Delta\tilde{T}_2^2 - \Delta\tilde{T}_1^2\right)}{s^2(x)}\left(\begin{bmatrix} 1 \\ 1 \end{bmatrix}(V_{1,2}^j)^T + V_{1,2}^j[1\ 1]\right).
\end{aligned}
\tag{B.46}
$$

The computation of the functional derivative relative to the term δE_3,

$$\delta E_3(x) \doteq -2\alpha_0 \int_\Omega \nabla s(x) \cdot \nabla h_s(x)dx \tag{B.47}$$

yields only one term different from 0

$$\nabla_s E_3(x) = -2\alpha_0 \triangle s(x); \tag{B.48}$$

similarly, the computation of the functional derivative relative to the term δE_4,

$$\delta E_4(x) \doteq 2\alpha_1 \int_\Omega \left(\int_\Omega s(y)dy - s^{mean}\right)h_s(x)dx \tag{B.49}$$

yields only one term different from 0,

$$\nabla_s E_4(x) = 2\alpha_1 \left(\int_\Omega s(y)dy - s^{mean}\right) \tag{B.50}$$

and also δE_5,

$$\delta E_5(x) \doteq 2\alpha_2 \int_{\partial\Omega_1} \frac{\nabla\phi(x)}{\|\nabla\phi(x)\|} \cdot \nabla h_\phi(x)dx \tag{B.51}$$

yields only one term different from 0,

$$\nabla_\phi E_5(x) = -2\alpha_2\delta(\phi(x))\nabla \cdot \left(\frac{\nabla\phi(x)}{\|\nabla\phi(x)\|}\right). \tag{B.52}$$

In the following, we report the more involved computation of the term δE_1,

$$\delta E_1 \doteq 2\int H(\phi(x))(u_1(x,t_1) - J(x,\Delta\tilde{T}_2))\delta u_1(x,t_1)\Big|_{h_r,h_s,h_{V_1}} dx \tag{B.53}$$

and skip the computation of

$$\delta E_2 \doteq 2\int H(-\phi(x))(u_2(x,t_1) - J(x,\Delta\tilde{T}_2))\delta u_2(x,t_1)\Big|_{h_r,h_s,h_{V_2}} dx \tag{B.54}$$

as it closely follows from that of δE_1. Again, to compute this functional derivative, we introduce the adjoint system

$$
\begin{cases}
\dot{w}_j(x,t) = \nabla \cdot (D_j(x)\nabla w_i(x,t)) \\
w_j(x,0) = u_j(x,t_1) - J(x,\Delta\tilde{T}_2) \\
(D_j(x)\nabla w_j(x,t)) \cdot n_j(x) = 0 \qquad \forall x \in \partial\Omega_j
\end{cases}
\tag{B.55}
$$

as we did in the previous section. Using Gauss' theorem, we now obtain

$$
\begin{aligned}
\delta E_1 &= 2 \int H(\phi(x)) w_1(x,0) \delta u_1(x,t_1)\, dx \\
&= 2 \int_0^{t_1}\!\!\int H(\phi(x))(\dot w_1(x,t_1-t)\delta u_1(x,t) \\
&\quad + w_1(x,t_1-t)\delta \dot u_1(x,t))\, dx\, dt \\
&= -2 \int_0^{t_1}\!\!\int H(\phi(x))(\delta D_1(x)\nabla u_1(x,t)) \cdot \nabla w_1(x,t_1-t)\, dx\, dt \\
&= \int \nabla_s E_1(x) h_s(x) + \nabla_\phi E_1(x) h_\phi(x)\, dx + \nabla_{V_{1,2}^1} E_1^T h_{V^1}.
\end{aligned}
$$

$$(B.56)$$

Using the formula for $\delta D_j(x)$ in equation (B.45) we can compute the gradients of equation (8.16) as

$$
\begin{aligned}
\nabla_s E(x) &= -2H(\phi(x)) \int_0^{t_1} (B_1(x)\nabla u_1(x,t)) \cdot \nabla w_1(x,t_1-t)\, dt \\
&\quad - 2H(-\phi(x)) \int_0^{t_1} (B_2(x)\nabla u_2(x,t)) \cdot \nabla w_2(x,t_1-t)\, dt \\
&\quad + 2\alpha_1 \left(\int_\Omega s(y)dy - s^{mean} \right) - 2\alpha_0 \triangle s(x) \\
\nabla_{V_{1,2}^1} E &= -2 \int_0^{t_1}\!\!\int_{\Omega_1} (C_1(x)\nabla u_1(x,t)) \cdot \nabla w_1(x,t_1-t)\, dx\, dt \\
\nabla_{V_{1,2}^2} E &= -2 \int_0^{t_1}\!\!\int_{\Omega\backslash\Omega_1} (C_2(x)\nabla u_2(x,t)) \cdot \nabla w_2(x,t_1-t)\, dx\, dt \\
\nabla_\phi E(x) &= \delta(\phi(x))(F_1(x) - F_2(x)) - 2\alpha_2\delta(\phi(x))\nabla \cdot \left(\frac{\nabla\phi(x)}{\|\nabla\phi(x)\|} \right),
\end{aligned}
$$

$$(B.57)$$

where for $j = 1,2$,

$$
F_j(x) = (u_j(x,t_1) - J(x,\Delta\tilde T_2))^2.
$$

$$(B.58)$$

Gradients of equation (8.41)

The first-order variation of the objective functional (8.41) along the direction $(h_s, h_{V_{1,2}^1}, h_{V_{1,2}^2}, h_b)$ is given by

$$
\begin{aligned}
\delta E &= 2 \int_\Omega (u_1(x,t_1) - u_2(x,t_1))(\delta u_1(x,t_1) - \delta u_2(x,t_1))\, dx \\
&\quad + 2\alpha_0 \int_\Omega \nabla s(x) \cdot \nabla h_s(x)dx \\
&\quad + 2\alpha_1 \int_\Omega \left(\int_\Omega s(y)dy - s^{mean} \right) h_s(x)dx,
\end{aligned}
$$

$$(B.59)$$

where

$$
\begin{aligned}
\hat{w}^1_{V^1}(x) &\doteq \frac{v}{s(x)}\sqrt{b^2+1}\Delta\tilde{T}_1 \\
\hat{w}^2_{V^1}(x) &\doteq \frac{v}{s(x)}b\Delta\tilde{T}_1 \\
\hat{w}^1_{V^2}(x) &\doteq \frac{v}{s(x)}b\Delta\tilde{T}_2 \\
\hat{w}^2_{V^2}(x) &\doteq \frac{v}{s(x)}\sqrt{b^2+1}\Delta\tilde{T}_2
\end{aligned}
\tag{B.60}
$$

and, similarly to the previous section, δu_j, $j = 1,2$ is the solution of the initial value problem

$$
\begin{cases}
\delta\dot{u}_j(x,t) = \nabla\cdot\left(\hat{D}_j(x)\nabla\delta u_j(x,t) + \delta\hat{D}_j(x)\nabla u_j(x,t)\right) \\
\delta u_j(x,0) = 0 & t \in (0,\infty) \\
(\hat{D}_j(x)\nabla\delta u_j(x,t) + \delta\hat{D}_j(x)\nabla u_j(x,t))\cdot n(x) = 0 & \forall x \in \partial\Omega
\end{cases}
\tag{B.61}
$$

with diffusion tensor $\hat{D}_j$ given by equation (8.28) and $\delta\hat{D}_j$, the first-order variation of the diffusion tensor, given by

$$
\begin{aligned}
\delta\hat{D}_j(x) &= -2\frac{\hat{W}^j(x)(\hat{W}^j(x))^T}{s(x)}h_s(x) \\
&+ \frac{v}{s(x)}\left(\left(\hat{W}^j(x)\hat{w}^j_{V^1}(x)\right)h_{V^1}^T + h_{V^1}\left(\hat{W}^j(x)\hat{w}^j_{V^1}(x)\right)^T\right) \\
&+ \frac{v}{s(x)}\left(\left(\hat{W}^j(x)\hat{w}^j_{V^2}(x)\right)h_{V^2}^T + h_{V^2}\left(\hat{W}^j(x)\hat{w}^j_{V^2}(x)\right)^T\right) \\
&+ \frac{1}{\sqrt{b^2+1}}\left(\hat{W}^j(x)(\hat{W}^{3-j}(x))^T + \hat{W}^{3-j}(x)(\hat{W}^j(x))^T\right)h_b.
\end{aligned}
\tag{B.62}
$$

The computation of the functional derivative relative to the term δE_2,

$$
\delta E_2 \doteq 2\alpha_0\int_\Omega \nabla s(x)\cdot\nabla h_s(x)dx
\tag{B.63}
$$

yields

$$
\nabla_s E_2(x) = -2\alpha_0\Delta s(x);
\tag{B.64}
$$

the computation of the functional derivative relative to the term δE_3,

$$
\delta E_3 \doteq 2\alpha_1\int_\Omega\left(\int_\Omega s(y)dy - s^{mean}\right)h_s(x)dx
\tag{B.65}
$$

yields

$$
\nabla_s E_3(x) = 2\alpha_1\left(\int_\Omega s(y)dy - s^{mean}\right);
\tag{B.66}
$$

the computation of the functional derivative relative to the term δE_4,

$$\delta E_4 \doteq -2\alpha_2 \int_\Omega \frac{\|\hat{W}^1(x)\|^2 + \|\hat{W}^2(x)\|^2}{s(x)} h_s(x) dx \qquad (B.67)$$

yields

$$\nabla_s E_4(x) = -2\alpha_2 \frac{\|\hat{W}^1(x)\|^2 + \|\hat{W}^2(x)\|^2}{s(x)}; \qquad (B.68)$$

the computation of the functional derivative relative to the terms δE_5 and δE_6,

$$
\begin{aligned}
\delta E_5 &\doteq 2\alpha_2 \left(\int_\Omega \hat{W}^1(x)\hat{w}_{V^1}^1(x) + \hat{W}^2(x)\hat{w}_{V^1}^2(x) dx \right)^T h_{V^1} \\
\delta E_6 &\doteq 2\alpha_2 \left(\int_\Omega \hat{W}^1(x)\hat{w}_{V^2}^1(x) + \hat{W}^2(x)\hat{w}_{V^2}^2(x) dx \right)^T h_{V^2}
\end{aligned}
\qquad (B.69)
$$

yields

$$
\begin{aligned}
\nabla_{V^1} E_5 &= 2\alpha_2 \int_\Omega \hat{W}^1(x)\hat{w}_{V^1}^1(x) + \hat{W}^2(x)\hat{w}_{V^1}^2(x) dx \\
\nabla_{V^2} E_6 &= 2\alpha_2 \int_\Omega \hat{W}^1(x)\hat{w}_{V^2}^1(x) + \hat{W}^2(x)\hat{w}_{V^2}^2(x) dx;
\end{aligned}
\qquad (B.70)
$$

and the computation of the functional derivative relative to the term δE_7,

$$\delta E_7 \doteq 2\alpha_2 \int_\Omega \frac{(\hat{W}^1(x))^T \hat{W}^2(x) + (\hat{W}^2(x))^T \hat{W}^1(x)}{\sqrt{b^2+1}} h_b \, dx \qquad (B.71)$$

yields

$$\nabla_b E_7 = 2\alpha_2 \int_\Omega \frac{(\hat{W}^1(x))^T \hat{W}^2(x) + (\hat{W}^2(x))^T \hat{W}^1(x)}{\sqrt{b^2+1}} dx. \qquad (B.72)$$

Now, we are left with the computation of the term δE_1,

$$\delta E_1 \doteq 2 \int_\Omega (u_1(x,t_1) - u_2(x,t_1))(\delta u_1(x,t_1) - \delta u_2(x,t_1)) \, dx. \qquad (B.73)$$

Similarly to the previous section, the adjoint problem is given by

$$
\begin{cases}
\dot{w}_j(x,t) = \nabla \cdot \left(\hat{D}_j(x)\nabla w_i(x,t) \right) \\
w_j(x,0) = \begin{cases} u_1(x,t_1) - u_2(x,t_1) & \text{if } j = 1 \\ u_2(x,t_1) - u_1(x,t_1) & \text{if } j = 2 \end{cases} \\
(\hat{D}_j(x)\nabla w_j(x,t)) \cdot n(x) = 0.
\end{cases}
\qquad (B.74)
$$

Using Gauss' theorem, we now obtain

$$
\begin{aligned}
\delta E_1 &= 2 \int_\Omega w_1(x,0)\delta u_1(x,t_1) \, dx + 2 \int_\Omega w_2(x,0)\delta u_2(x,t_1) \, dx \\
&= \int_\Omega \nabla_s E_1(x)h_s(x)dx + \nabla_{V^1} E_1^T h_{V^1} + \nabla_{V^2} E_1^T h_{V^2} + \nabla_b E_1 h_b.
\end{aligned}
\qquad (B.75)
$$

Using the formula for $\delta \hat{D}_j(x)$ in equation (B.62) we can compute the gradients of equation (8.41) as

$$
\begin{aligned}
\nabla_s E(x) &= -2 \int_0^{t_1} \left(\frac{\hat{W}^1(x)(\hat{W}^1(x))^T}{s(x)} \nabla u_1(x,t) \right) \cdot \nabla w_1(x, t_1 - t)\, dt \\
&\quad - 2 \int_0^{t_1} \left(\frac{\hat{W}^2(x)(\hat{W}^2)^T(x)}{s(x)} \nabla u_2(x,t) \right) \cdot \nabla w_2(x, t_1 - t)\, dt \\
&\quad - 2\alpha_0 \triangle s(x) + 2\alpha_1 \left(\int_\Omega s(y)dy - s^{mean} \right) - \\
&\quad - 2\alpha_2 \frac{\|\hat{W}^1(x)\|^2 + \|\hat{W}^2(x)\|^2}{s(x)} \\
\nabla_{V^1} E &= -2 \int_0^{t_1} \int_\Omega \left(B_1^1(x)\nabla u_1(x,t) \right) \cdot \nabla w_1(x, t_1 - t)\, dx\, dt \\
&\quad - 2 \int_0^{t_1} \int_\Omega \left(B_1^2(x)\nabla u_2(x,t) \right) \cdot \nabla w_2(x, t_1 - t)\, dx\, dt \\
&\quad + 2\alpha_2 \int_\Omega \hat{W}^1(x)\hat{w}_{V^1}^1(x) + \hat{W}^2(x)\hat{w}_{V^1}^2(x)dx \\
\nabla_{V^2} E &= -2 \int_0^{t_1} \int_\Omega \left(B_2^1(x)\nabla u_1(x,t) \right) \cdot \nabla w_1(x, t_1 - t)\, dx\, dt \\
&\quad - 2 \int_0^{t_1} \int_\Omega \left(B_2^2(x)\nabla u_2(x,t) \right) \cdot \nabla w_2(x, t_1 - t)\, dx\, dt \\
&\quad + 2\alpha_2 \int_\Omega \hat{W}^1(x)\hat{w}_{V^2}^1(x) + \hat{W}^2(x)\hat{w}_{V^2}^2(x)dx \\
\nabla_b E &= -2 \int_0^{t_1} \int_\Omega \left(F(x)\nabla u_1(x,t) \right) \cdot \nabla w_1(x, t_1 - t)\, dx\, dt \\
&\quad - 2 \int_0^{t_1} \int_\Omega \left(F(x)\nabla u_2(x,t) \right) \cdot \nabla w_2(x, t_1 - t)\, dx\, dt \\
&\quad + 2\alpha_2 \int_\Omega \frac{(\hat{W}^1(x))^T \hat{W}^2(x) + (\hat{W}^2(x))^T \hat{W}^1(x)}{\sqrt{b^2 + 1}} dx.
\end{aligned}
$$

$$(B.76)$$

where for $i = 1, 2$ and $j = 1, 2$,

$$
\begin{aligned}
B_i^j(x) &= \frac{v}{s(x)} \left(\hat{W}^j(x)\hat{w}_{V^i}^j(x) [1\ 1] + \begin{bmatrix} 1 \\ 1 \end{bmatrix} \left(\hat{W}^j(x)\hat{w}_{V^i}^j(x) \right)^T \right) \\
F(x) &= \frac{1}{\sqrt{b^2 + 1}} \left(\hat{W}^1(x)(\hat{W}^2(x))^T + \hat{W}^2(x)(\hat{W}^1(x))^T \right).
\end{aligned}
$$

$$(B.77)$$

Gradients of equation (8.61)

We formally compute directional derivatives in direction $(h_r, h_s, h_{V^{i,1}}, h_{V^{i,2}}, h_\phi)$ for $i = 1, \ldots, K$ in the following. The first-order variation of the objective functional (8.61) is given by

$$
\begin{aligned}
\delta E \;=\;& 2\sum_{i=1}^{K}\int H(\phi(x))(u_{i,1}(x,t_1) - J(x,\Delta\tilde{T}_i))\delta u_{i,1}(x,t_1)dx \\
&+ 2\sum_{i=1}^{K}\int H(-\phi(x))(u_{i,2}(x,t_1) - J(x,\Delta\tilde{T}_i))\delta u_{i,2}(x,t_1)dx \\
&+ 2\alpha_0 \int_\Omega (r(x) - r^*(x))\, h_r(x)dx \\
&+ 2\alpha_1 \int_\Omega \nabla s(x)\cdot\nabla h_s(x)dx \\
&+ 2\alpha_2 \int_\Omega \left(\int_\Omega s(y)dy - s^{mean}\right)h_s(x)dx \\
&- 2\alpha_3 \int_\Omega \delta(\phi(x))\nabla\cdot\left(\frac{\nabla\phi}{\|\nabla\phi(x)\|}\right)h_\phi(x)dx \\
&+ 2\sum_{i=1}^{K}\int \delta(\phi(x))(u_{i,1}(x,t_1) - J(x,\Delta\tilde{T}_i))^2 h_\phi(x)dx \\
&- 2\sum_{i=1}^{K}\int \delta(\phi(x))(u_{i,2}(x,t_1) - J(x,\Delta\tilde{T}_i))^2 h_\phi(x)dx,
\end{aligned}
$$

$$(B.78)$$

where, similarly to the previous section, $\delta u_{i,j}$ is the solution of the initial value problem

$$
\begin{cases}
\delta\dot{u}_{i,j}(x,t) = \nabla\cdot(D_{i,j}(x)\nabla\delta u_{i,j}(x,t) + \delta D_{i,j}(x)\nabla u_{i,j}(x,t)) \\
\delta u_{i,j}(x,0) = h_r(x) \\
(D_{i,j}(x)\nabla\delta u_{i,j}(x,t) + \delta D_{i,j}(x)\nabla u_{i,j}(x,t))\cdot n_j(x) = 0 \qquad \forall x\in\partial\Omega_j
\end{cases}
$$

$$(B.79)$$

and $\delta D_{i,j}$ is the variation of the diffusion tensor, given by

$$
\begin{aligned}
\delta D_{i,j}(x) \;=\;& 2\left(\frac{\sigma^2(x)}{s^2(x)\left(\frac{1}{F}-\frac{1}{v}\right)-s(x)}I_d - \frac{v^2\Delta\tilde{T}_i^2}{s^3(x)}V_{1,2}^{i,j}(V_{1,2}^{i,j})^T\right)h_s(x) \\
&+ \frac{v^2\Delta\tilde{T}_i^2}{s^2(x)}\left(h_{V^{i,j}}(V_{1,2}^{i,j})^T + V_{1,2}^{i,j}h_{V^{i,j}}^T\right) \qquad \forall x\in\Omega_j.
\end{aligned}
$$

$$(B.80)$$

The computation of the functional derivative relative to the term δE_3,

$$
\delta E_3 \doteq 2\alpha_0 \int_\Omega (r(x) - r^*(x))\, h_r(x)dx \tag{B.81}
$$

yields only one term different from 0,

$$
\nabla_r E_3(x) = 2\alpha_0\,(r(x) - r^*(x)); \tag{B.82}
$$

the computation of the functional derivative relative to the term δE_4,

$$
\delta E_4 \doteq 2\alpha_1 \int_\Omega \nabla s(x)\cdot\nabla h_s(x)dx \tag{B.83}
$$

yields only one term different from 0,

$$\nabla_s E_4(x) = -2\alpha_1 \triangle s(x); \tag{B.84}$$

the computation of the functional derivative relative to the term δE_5,

$$\delta E_5 \doteq 2\alpha_2 \int_\Omega \left(\int_\Omega s(y)dy - s^{mean} \right) h_s(x)dx \tag{B.85}$$

yields only one term different from 0,

$$\nabla_s E_5(x) = 2\alpha_2 \left(\int_\Omega s(y)dy - s^{mean} \right) \tag{B.86}$$

and also δE_6,

$$\begin{aligned}
\delta E_6 \doteq\ & -2\alpha_3 \int_\Omega \delta(\phi(x))\nabla \cdot \left(\frac{\nabla\phi(x)}{\|\nabla\phi(x)\|} \right) h_\phi(x)dx \\
& + 2\sum_{i=1}^{K} \int \delta(\phi(x))(u_{i,1}(x,t_1) - J(x,\Delta\tilde{T}_i))^2 h_\phi(x)dx \\
& - 2\sum_{i=1}^{K} \int \delta(\phi(x))(u_{i,2}'(x,t_1) - J(x,\Delta\tilde{T}_i))^2 h_\phi(x)dx
\end{aligned} \tag{B.87}$$

yields

$$\begin{aligned}
\nabla_\phi E_6(x) =\ & -2\alpha_3\delta(\phi(x))\nabla \cdot \left(\frac{\nabla\phi}{\|\nabla\phi(x)\|} \right) \\
& + 2\sum_{i=1}^{K} \delta(\phi(x))(u_{i,1}(x,t_1) - J(x,\Delta\tilde{T}_i))^2 \\
& - 2\sum_{i=1}^{K} \delta(\phi(x))(u_{i,2}(x,t_1) - J(x,\Delta\tilde{T}_i))^2.
\end{aligned} \tag{B.88}$$

In the following, we report the computation of the term $\delta E_{i,1}$,

$$\delta E_{i,1} \doteq 2\int H(\phi(x))(u_{i,1}(x,t_1) - J(x,\Delta\tilde{T}_i))\delta u_{i,1}(x,t_1)dx \tag{B.89}$$

and skip the computation of

$$\delta E_{i,2} \doteq 2\int H(-\phi(x))(u_{i,2}(x,t_1) - J(x,\Delta\tilde{T}_i))\delta u_{i,2}(x,t_1)dx \tag{B.90}$$

as it closely follows from that of $\delta E_{i,1}$. Again we employ the adjoint method by introducing the following system,

$$\begin{cases}
\dot{w}_{i,j}(x,t) = \nabla \cdot (D_{i,j}(x)\nabla w_{i,j}(x,t)) \\
w_{i,j}(x,0) = u_{i,j}(x,t_1) - J(x,\Delta\tilde{T}_i) \\
(D_{i,j}(x)\nabla w_{i,j}(x,t)) \cdot n_j(x) = 0 \qquad \forall x \in \partial\Omega_j.
\end{cases} \tag{B.91}$$

As we have done in the previous sections, we start by writing

$$\delta E_{i,1} = 2\int H(\phi(x))w_{i,1}(x,0)\delta u_{i,1}(x,t_1)\, dx \tag{B.92}$$

and rearrange all the terms so that the first-order variation of $E_{i,1}$ can be rewritten as

$$
\delta E_{i,1} = \int \left(\nabla_r E_{i,1}(x) h_r(x) + \nabla_s E_{i,1}(x) h_s(x) \right. \\
\left. + \nabla_\phi E_{i,1}(x) h_\phi(x) \right) dx + \nabla_{V^{i,1}} E_{i,1}^T h_{V^{i,1}}. \tag{B.93}
$$

Using the formula for $\delta D_{i,j}(x)$ in equation (B.80) we can compute the gradients of equation (8.61):

$$
\begin{aligned}
\nabla_r E(x) &= 2 \sum_{i=1}^{K} \left(H(\phi(x)) w_{i,1}(x, t_1) + H(-\phi(x)) w_{i,2}(x, t_1) \right) \\
&\quad + 2\alpha_0 (r(x) - r^*(x)) \\
\nabla_s E(x) &= -2 \sum_{i=1}^{K} \int_0^{t_1} (B_{i,1}(x) \nabla u_{i,1}(x,t)) \cdot \nabla w_{i,1}(x, t_1 - t) \, dt \\
&\quad - 2 \sum_{i=1}^{K} \int_0^{t_1} (B_{i,2}(x) \nabla u_{i,2}(x,t)) \cdot \nabla w_{i,2}(x, t_1 - t) \, dt \\
&\quad - 2\alpha_1 \Delta s(x) + 2\alpha_2 \left(\int_\Omega s(y) dy - s^{mean} \right) \\
\nabla_{V^{i,1}} E &= -2 \int_0^{t_1} \int (C_{i,1}(x) \nabla u_{i,1}(x,t)) \cdot \nabla w_{i,1}(x, t_1 - t) \, dx \, dt \\
\nabla_{V^{i,2}} E &= -2 \int_0^{t_1} \int (C_{i,2}(x) \nabla u_{i,2}(x,t)) \cdot \nabla w_{i,2}(x, t_1 - t) \, dx \, dt \\
\nabla_\phi E(x) &= \delta(\phi(x)) \left(\sum_{i=1}^{K} (F_{i,1}(x) - F_{i,2}(x)) - 2\alpha_3 \nabla \cdot \left(\frac{\nabla \phi(x)}{\|\nabla \phi(x)\|} \right) \right),
\end{aligned} \tag{B.94}
$$

where

$$
\begin{aligned}
B_{i,1}(x) &= 2H(\phi(x)) \left(\frac{\sigma^2(x)}{s^2(x) \left(\frac{1}{F} - \frac{1}{v} \right) - s(x)} I_d - \frac{v^2 \Delta \tilde{T}_i^2}{s^3(x)} V_{1,2}^{i,1} (V_{1,2}^{i,1})^T \right) \\
B_{i,2}(x) &= 2H(-\phi(x)) \left(\frac{\sigma^2(x)}{s^2(x) \left(\frac{1}{F} - \frac{1}{v} \right) - s(x)} I_d - \frac{v^2 \Delta \tilde{T}_i^2}{s^3(x)} V_{1,2}^{i,2} (V_{1,2}^{i,2})^T \right) \\
C_{i,1}(x) &= 2H(\phi(x)) \frac{v^2 \Delta \tilde{T}_i^2}{s^2(x)} \left(\begin{bmatrix} 1 \\ 1 \end{bmatrix} (V_{1,2}^{i,1})^T + V_{1,2}^{i,1} [1 \ 1] \right) \\
C_{i,2}(x) &= 2H(-\phi(x)) \frac{v^2 \Delta \tilde{T}_i^2}{s^2(x)} \left(\begin{bmatrix} 1 \\ 1 \end{bmatrix} (V_{1,2}^{i,2})^T + V_{1,2}^{i,2} [1 \ 1] \right) \\
F_{i,1}(x) &= (u_{i,1}(x, t_1) - J(x, \Delta \tilde{T}_1))^2 \\
F_{i,2}(x) &= (u_{i,2}(x, t_1) - J(x, \Delta \tilde{T}_2))^2.
\end{aligned} \tag{B.95}
$$

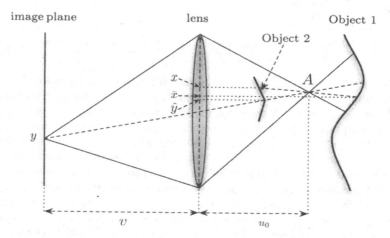

Figure B.1. Geometry of a scene with occlusions. The coordinates y on the image plane are projected on surface s_2 through the center of the lens, and become the coordinates $\hat{y}$. Similarly, the coordinate x on the surface s_1 is projected to the surface s_2 through the point A of coordinates $(-yu_0/v, u_0)$, and becomes the coordinate $\bar{x}$.

B.2.4 Computation of the gradients in Chapter 9

The computation of the gradients of the cost functional (9.6) is not straightforward. The hidden complexity lies in the image formation model. Let us recall equation (2.27):

$$I(y) = \int_{\Omega_1} h_{s_1}(y,x) r_1(x) dx + \int_{\Omega_2} h_{s_2}(y,x) r_2(x) dx. \tag{B.96}$$

The main challenge in computing this expression is in the evaluation of the domains Ω_1 and Ω_2. Because we used the level set representation to model both domains, we can write

$$\Omega_1 = \{x | \phi(x) > 0\} \tag{B.97}$$

and

$$\Omega_2 = \{x | \phi(\varpi(x)) < 0\}, \tag{B.98}$$

where ϖ is the projection of a point on the first surface s_1 onto the surface s_2 through the point A as shown in Figure B.1. Although the evaluation of Ω_1 does not present particular challenges, the evaluation of Ω_2 is much more involved. The source of most problems is the projection ϖ. First of all notice that the projection depends on the coordinate y on the image domain. This coordinate indeed determines the center of projection $A = (-yu_0/v, u_0)$. Unfortunately, this is not the only difficulty. To see where the other problems lie, let us write the projection

ϖ explicitly:

$$\varpi(x) \doteq \bar{x} = y\frac{u_0}{v}\frac{s_1(x) - s_2(\bar{x})}{s_1(x) - u_0} + x\frac{s_2(\bar{x}) - u_0}{s_1(x) - u_0}, \tag{B.99}$$

where u_0 is the parameter u that satisfies the thin lens law in equation (2.1) with focus setting v. Notice that equation (B.99) depends implicitly on $\bar{x}$ via the surface s_2 and, similarly, it depends implicitly on x via the surface s_1, therefore, as is, it is not suitable to determine the domain Ω_2. Furthermore, as we show, we also need to be able to compute the inverse projection ϖ^{-1} in the numerical integration of equation (B.96). Before proceeding with our solution to the computation of the projection ϖ, we want to point out a more subtle issue. The two radiances r_1 and r_2 are eventually represented with a finite set of parameters. For instance, one could choose a grid parameter of the same size as the input images. Notice, however, that we need to take the chosen grid into account during the numerical integration of the radiances against the respective kernels in equation (B.96). Depending on the projection ϖ and the relative distances between the surfaces s_1 and s_2, the grid of the radiance r_2 is magnified or reduced. This also contributes to increase the complexity of the numerical integration.

We now show how to simplify both of these problems by approximating the projection ϖ and reparameterizing all the variables with respect to a common domain. Let us start with the projection ϖ. As mentioned above, the main challenge is that equation (B.99) is recursive in $\bar{x}$. To solve this problem we expand the surface s_2 in Taylor series around the point $\hat{y}$; that is,

$$s_2(\bar{x}) = s_2(\hat{y}) + \nabla s_2^T(\hat{y})(\bar{x} - \hat{y}) + o(\|\bar{x} - \hat{y}\|^2), \tag{B.100}$$

where ∇s_2 is the gradient of the surface s_2 and $o(\|\bar{x} - \hat{y}\|^2)$ denotes terms that are nonlinear in $\bar{x}$. If we consider only the first term in the expansion, that is, $s_2(\bar{x}) = s_2(\hat{y})$, then we are approximating the surface s_2 with a fronto-parallel plane. Of course, one can also consider more terms in the expansion and then try to find an explicit solution in $\bar{x}$. For instance, by considering the expansion up to the linear term, we obtain:

$$\varpi(x) = \left(I_d - \frac{\left(x - y\frac{u_0}{v}\right)\nabla s_2(\hat{y})^T}{s_1(x) - u_0}\right)^{-1}\left(x\frac{s_2(\hat{y}) + \nabla s_2(\hat{y})^T\hat{y} - u_0}{s_1(x) - u_0}\right.$$
$$\left. + y\frac{u_0}{v}\frac{s_1(x) - s_2(\hat{y}) - \nabla s_2(\hat{y})^T\hat{y}}{s_1(x) - u_0}\right); \tag{B.101}$$

however, this solution is still intractable because we also need the inverse projection ϖ^{-1}. Hence, for simplicity, we consider only the fronto-parallel approximation for both surfaces and obtain:

$$\varpi(x) \simeq -\hat{y}\frac{u_0}{s_2(\hat{y})}\frac{s_1(\hat{y}) - s_2(\hat{y})}{s_1(\hat{y}) - u_0} + x\frac{s_2(\hat{y}) - u_0}{s_1(\hat{y}) - u_0}, \tag{B.102}$$

which is a linear change of coordinates in x and therefore more tractable when computing the numerical integration of the image formation model (B.96). To compute the inverse projection, we can also write

$$\varpi^{-1}(\bar{x}) \doteq x \simeq \hat{y}\frac{u_0}{s_2(\hat{y})}\frac{s_1(\hat{y}) - s_2(\hat{y})}{s_2(\hat{y}) - u_0} + \bar{x}\frac{s_1(\hat{y}) - u_0}{s_2(\hat{y}) - u_0}. \tag{B.103}$$

As mentioned above, this approximation yields the exact projection when both surfaces are fronto-parallel planes, and the zeroth order solution otherwise.

Now, because we have knowledge of the focus settings $v_1 \ldots v_K$, we further simplify our computations by rescaling the coordinates between images $I(\cdot, v_1) \ldots I(\cdot, v_K)$ with respect to a reference focus setting that we denote with v:

$$y = -\frac{\hat{y}v}{s_2(\hat{y})} \tag{B.104}$$

so that the change of coordinates does not change with v_i. Rescaling is done by using an offline calibration procedure (see Appendix D), but can also be avoided by using telecentric optics [Watanabe and Nayar, 1996b]. Furthermore, to simplify notation, we introduce $\bar{I}(\cdot, v_i)$, the projection of the input images onto the surface s_2, by defining

$$\bar{I}(\hat{y}, v_i) = I\left(-\frac{\hat{y}v}{s_2(\hat{y})}, v_i\right) = I(y, v_i). \tag{B.105}$$

Notice that given the coordinate $\hat{y}$ it is immediate to find the coordinate y (given the surface s_2), but it is not immediate to find $\hat{y}$ given the coordinate y. Finally, let us define the projection $\bar{r}_1$ of the radiance r_1 on the surface s_2; that is,

$$\bar{r}_1(\bar{x}) = r_1\left(\varpi^{-1}(\bar{x})\right). \tag{B.106}$$

With these considerations at hand, we redefine the imaging model equation (2.27) as

$$\begin{aligned}\bar{I}(\hat{y}, v_i) &= \int_{\mathbb{R}^2} H\big(\phi(\bar{x})\big)h_{s_2(\hat{y})}^{v_i}(\hat{y}, \bar{x})r_2(\bar{x})d\bar{x} + \int_{\mathbb{R}^2}\big(1 - H\big(\phi(\bar{x})\big)\big)\\ &\quad h_{s_1(\hat{y})}^{v_i}\left(\hat{y}\frac{s_1(\hat{y})}{s_2(\hat{y})}, \varpi^{-1}(\bar{x})\right)\bar{r}_1(\bar{x})\frac{s_1(\hat{y}) - u_0}{s_2(\hat{y}) - u_0}d\bar{x}\end{aligned} \tag{B.107}$$

and notice that the two integrals are now defined on the same domain of integration but are nonzero on complementary regions. The imaging model above is then introduced in the cost functional (9.6) and yields:

$$\begin{aligned}E &= \sum_{i=1}^{K}\int_\Omega\big(\bar{I}(\hat{y}, v_i) - \bar{J}(\hat{y}, v_i)\big)^2 d\hat{y}\\ &\quad + \alpha_1\|r_1 - r_1^*\|^2 + \alpha_2\|r_2 - r_2^*\|^2\\ &\quad + \alpha_3\|\nabla s_1\|^2 + \alpha_4\|\nabla s_2\|^2\\ &\quad + \alpha_5\|\nabla H(\phi)\|^2.\end{aligned} \tag{B.108}$$

To minimize (B.108), we need to compute the first-order variations of the cost functional with respect to the five unknowns $\{\bar{r}_1, r_2, s_1, s_2, \phi\}$. The first gradient we examine is $\nabla_{\bar{r}_1} E$, and it is composed of two terms:

$$
\begin{aligned}
\nabla_{\bar{r}_1} E(\bar{x}) \;=\; & 2\left(1 - H\left(\phi(\bar{x})\right)\right) \sum_{i=1}^{K} \int \left(\bar{I}(\hat{y}, v_i) - \bar{J}(\hat{y}, v_i)\right) \frac{s_1(\hat{y}) - u_0}{s_2(\hat{y}) - u_0} \\
& h^{v_i}_{s_1(\hat{y})} \left(\hat{y}\frac{s_1(\hat{y})}{s_2(\hat{y})}, \varpi^{-1}(\bar{x})\right) \, d\hat{y} + 2\alpha_1(\bar{r}_1(\bar{x}) - r_1^*(\bar{x})).
\end{aligned}
$$
(B.109)

In equation (B.109) the first term comes from "matching" the image model and the measured image, and the other term comes from the regularization imposed on $\bar{r}_1$. Similarly, we obtain the gradient $\nabla_{r_2} E$:

$$
\begin{aligned}
\nabla_{r_2} E(\bar{x}) \;=\; & 2H\left(\phi(\bar{x})\right) \sum_{i=1}^{K} \int \left(\bar{I}(\hat{y}, v_i) - \bar{J}(\hat{y}, v_i)\right) h^{v_i}_{s_2(\hat{y})}(\hat{y}, \bar{x}) \, d\hat{y} \\
& + 2\alpha_2(r_2(\bar{x}) - r_2^*(\bar{x})).
\end{aligned}
$$
(B.110)

Due to the complexity of the analytic formula of the gradient $\nabla_{s_1} E$ we resort to the finite gradient computation. The finite gradient is a much simpler, although more computationally intensive, way to compute the gradient than the analytic derivation. Because the surface s_1 is parameterized with respect to the same coordinates of the images, we can approximate $\nabla_{s_1} E$ pointwise via

$$
\nabla_{s_1} E(\hat{y}) \simeq \frac{E(s_1(\hat{y}) + ds_1(\hat{y})) - E(s_1(\hat{y}))}{ds_1(\hat{y})}
$$
(B.111)

for a finite value $ds_1(\hat{y}) \ll s_1(\hat{y}) \; \forall \; \hat{y} \in \Omega$. In practice, due to the discretization of the functions in the implementation of the algorithm, it is sufficient to choose $ds_1(\hat{y})$ between 0.1% and 1% of s_1 to compute a reasonable approximation of $\nabla_{s_1} E$. Similarly, we approximate $\nabla_{s_2} E$ pointwise by

$$
\nabla_{s_2} E(\hat{y}) \simeq \frac{E(s_2(\hat{y}) + ds_2(\hat{y})) - E(s_2(\hat{y}))}{ds_2(\hat{y})}
$$
(B.112)

for a finite value $ds_2(\hat{y}) \ll s_2(\hat{y}) \; \forall \; \hat{y} \in \Omega$. The computation of the last gradient $\nabla_\phi E$ yields

$$
\begin{aligned}
\nabla_\phi E(\bar{x}) \;=\; & 2\delta(\phi(\bar{x})) \sum_{i=1}^{K} \int \left(\bar{I}(\hat{y}, v_i) - \bar{J}(\hat{y}, v_i)\right) \left(h^{v_i}_{s_2(\hat{y})}(\hat{y}, \bar{x}) r_2(\bar{x}) \right. \\
& \left. - h^{v_i}_{s_1(\hat{y})} \left(\hat{y}\frac{s_1(\hat{y})}{s_2(\hat{y})}, \varpi^{-1}(\bar{x})\right) \bar{r}_1(\bar{x}) \frac{s_1(\hat{y}) - u_0}{s_2(\hat{y}) - u_0} \right) d\hat{y} \\
& - 2\alpha_5 \delta(\phi(\bar{x})) \nabla \cdot \left(\frac{\nabla\phi(\bar{x})}{|\nabla\phi(\bar{x})|} \right).
\end{aligned}
$$
(B.113)

One may find the computation of the gradient of the Heaviside function, which is discontinuous, a bit uneasy. Because our formulation of the problem requires the computation of gradients, which are not well-defined for the Heaviside function[3] in the context of regular functions, we introduce the following regularized version [Chan and Vese, 2001] of the Heaviside function

$$H_\epsilon(z) \doteq \frac{1}{2} \left(1 + \frac{2}{\pi} \arctan \left(\frac{z}{\epsilon} \right) \right),$$
(B.114)

where ϵ is a tuning parameter and determines the degree of regularization of H_ϵ. In a similar fashion, we define the derivative of $H_\epsilon(z)$ as the regularized Dirac delta

$$\delta_\epsilon(z) \doteq \frac{\partial H_\epsilon(z)}{\partial z} = \frac{1}{\pi \epsilon} \frac{1}{1 + \left(\frac{z}{\epsilon} \right)^2}.$$
(B.115)

---

[3]The derivative of the Heaviside is, however, well defined in a distributional sense (see [McOwen, 1996]).

Appendix C
Proofs

C.1 Proof of Proposition 3.2

Proposition 3.2. *Any smooth surface s is weakly observable.*

Proof. We prove the statement by contradiction. Suppose there exists a surface s' and an open $O \subset \Omega$ such that $s'(x) \neq s(x)$ $\forall\, x \in O$. Then, from the definition of weak indistinguishability, we have that for any radiance r there exists a radiance r' such that $I_s^r(y, v) = I_{s'}^{r'}(y, v), \forall\, y \in \Omega\ \forall\, v \in \mathcal{V}$. By Property 1 in Section 3.2, for any $y \in \Omega$ there exists a focus setting $\bar{v}$ such that $I_s^r(y, \bar{v}) = r(y)$ and hence

$$I_{s'}^{r'}(y, \bar{v}) \doteq \int h_{s'}^{\bar{v}}(y, x) r'(x) dx = r(y). \tag{C.1}$$

Similarly, for any $y \in \Omega$ there exists a focus setting $\bar{v}'$ such that $I_{s'}^{r'}(y, \bar{v}') = r'(y)$ and

$$I_s^r(y, \bar{v}') \doteq \int h_s^{\bar{v}'}(y, x) r(x) dx = r'(y). \tag{C.2}$$

Notice that, for a given radiance r, we obtained an explicit expression of the radiance r' such that s and s' are indistinguishable. Substituting equation (C.2) into equation (C.1), we obtain

$$\iint h_{s'}^{\bar{v}}(y, x) h_s^{\bar{v}'}(x, \tilde{x}) r(\tilde{x}) d\tilde{x} dx = r(y) \qquad \forall\, y \in \Omega. \tag{C.3}$$

Because the above equation holds for any r, it follows that

$$\int h_{s'}^{\bar{v}}(y, x) h_s^{\bar{v}'}(x, \tilde{x}) dx = \delta(y - \tilde{x}) \qquad \forall\, y, \tilde{x} \in O. \tag{C.4}$$

However, because $s'(x) \neq s(x)$ for $x \in O \subset \Omega$, then $\bar{v}$ is not a focus setting for s' in O and $\bar{v}'$ is not a focus setting for s in O by Property 2. This means that, by Property 3, there exists an open set $\bar{O} \subset O$ such that $h_{s'}^{\bar{v}}(y, x) > 0$ and $h_s^{\bar{v}'}(x, \tilde{x}) > 0$ for any $y, x, \tilde{x} \in \bar{O}$. Hence, if we choose $y, \tilde{x} \in \bar{O}$ with $y \neq \tilde{x}$, we have

$$\int h_{s'}^{\bar{v}}(y, x) h_s^{\bar{v}'}(x, \tilde{x}) dx > 0 \neq \delta(y - \tilde{x}) = 0, \tag{C.5}$$

which is a contradiction.   □

C.2   Proof of Proposition 3.5

Proposition 3.5. *If* $r : \mathbb{R}^2 \mapsto \mathbb{R}$ *is harmonic, then, for any integrable function* $h : \mathbb{R}^2 \mapsto \mathbb{R}$ *that is rotationally symmetric,*

$$\int h(y - x) r(x) dx = r(y) \int h(x) dx \tag{C.6}$$

whenever the integrals exist.

Proof. Let $f(|x|) = F(x)$, where $f : [0, \infty) \mapsto \mathbb{R}$. Suppose that equation (C.6) is true in the case $y = 0$. Then it is true in any case; indeed, we could otherwise substitute $r_y(x) = r(x + y)$, and

$$\begin{aligned} r(y) \int F(x) dx &= r_y(0) \int F(x) dx = \int f(|\tilde{x}|) r_y(\tilde{x}) d\tilde{x} \\ &= \int f(|-\tilde{x}|) r(\tilde{x} + y) d\tilde{x} = \int f(|y - x|) r(x) dx. \end{aligned} \tag{C.7}$$

Now, we are left with proving the proposition for $y = 0$. We change coordinates by

$$\int f(|x|) r(x) \, dx = \int_0^\infty \int_{C_\rho} f(\rho) r(z) \, d\rho \, d\mu_\rho(z) = \int_0^\infty f(\rho) \int_{C_\rho} r(z) \, d\mu_\rho(z) \, d\rho, \tag{C.8}$$

where C_ρ is the circumference of radius ρ centered in 0, $z \in C_\rho$, and μ_ρ is the measure on C_ρ (e.g., $z = \rho[\cos(\theta), \sin(\theta)]^T$, $d\mu_\rho(z) = \rho d\theta$ and the integral is between 0 and 2π). Then, by the mean-value theorem, (see [Taylor, 1996], Proposition 2.4 in Section 3.2)

$$\int_{C_\rho} r(z) \, d\mu_\rho(z) = 2\pi\rho r(0)$$

and then

$$\int f(|x|) r(x) \, dx = r(0) \int_0^\infty f(\rho) 2\pi\rho \, d\rho = r(0) \int F(x) \, dx.$$

□

C.3   Proof of Proposition 4.1

Proposition 4.1. *Let $\hat{s}$, $\hat{r}$ be local extrema of the functional*

$$E(s,r) \doteq |J - H_s r|^2 \tag{C.9}$$

and, assuming that $H_s^\dagger$ exists, let $\tilde{s}$ be a local extremum of the functional

$$\psi(s) \doteq |H_s^\perp J|^2. \tag{C.10}$$

Furthermore, let $\tilde{r}$ be obtained from s by

$$\tilde{r} \doteq H_s^\dagger J. \tag{C.11}$$

Then $\hat{s}$ is also a local extremum of $\psi(s)$, and $\tilde{s}, \tilde{r}$ are also local extrema of $E(s,r)$.

The proof of the above proposition closely relates to the results of [Golub and Pereyra, 1973]. Before proceeding with the proof, we need to introduce some additional notation. The extrema of the functional E and ψ are defined via their Fréchet functional derivatives. They result in the following coupled equations

$$\begin{cases} \nabla_s E(\hat{s}, \hat{r}) = 0 \\ \nabla_r E(\hat{s}, \hat{r}) = 0 \end{cases} \tag{C.12}$$

and

$$\begin{cases} \nabla_s \psi(\tilde{s}) = 0 \\ r \doteq H_{\tilde{s}}^\dagger \end{cases}, \tag{C.13}$$

where $\nabla_s E$ and $\nabla_r E$ stand for the gradients of E with respect to s and r, respectively, and $\nabla_s \psi$ stands for the gradient of ψ with respect to s [Luenberger, 1968]. For simplicity, where it is possible, we indicate with $\dot{A}$ the derivative of A with respect to s instead of $\nabla_s A$.

For ease of reading, we simplify the proof of Proposition 4.1 by gathering some of the results in the following lemma.

Lemma C.1. *Let $P_{H_s} \doteq H_s H_s^\dagger$ be the projection operator onto the range of H_s and recall that $H_s^\dagger$ verifies $H_s H_s^\dagger H_s = H_s$ and $(H_s H_s^\dagger)^* = H_s H_s^\dagger$, then*

$$\dot{P}_{H_s} = H_s^\perp \dot{H}_s H_s^\dagger + \left(H_s^\perp \dot{H}_s H_s^\dagger \right)^*. \tag{C.14}$$

Proof. Because $P_{H_s} H_s = H_s$, then:

$$\nabla_s (P_{H_s} H_s) = \dot{P}_{H_s} H_s + P_{H_s} \dot{H}_s = \dot{H}_s \tag{C.15}$$

and

$$\dot{P}_{H_s} H_s = \dot{H}_s - P_{H_s} \dot{H}_s = H_s^\perp \dot{H}_s. \tag{C.16}$$

Also,

$$\dot{P}_{H_s} P_{H_s} = \dot{P}_{H_s} H_s H_s^\dagger = H_s^\perp \dot{H}_s H_s^\dagger. \tag{C.17}$$

Because $\left(\dot{P}_{H_s}P_{H_s}\right)^* = P_{H_s}\dot{P}_{H_s}$, then

$$\begin{aligned}\dot{P}_{H_s} &= \nabla_s(P_{H_s}P_{H_s}) = \dot{P}_{H_s}P_{H_s} + P_{H_s}\dot{P}_{H_s}\\ &= H_s^{\perp}\dot{H}_s H_s^{\dagger} + \left(H_s^{\perp}\dot{H}_s H_s^{\dagger}\right)^*\end{aligned} \qquad (\text{C.18})$$

which completes the proof. $\square$

We now use the results of Lemma C.1, together with the fact that $\dot{H}_s^{\perp} = -\dot{P}_{H_s}$, to prove Proposition 4.1.

Proof. We have that

$$\frac{1}{2}\nabla_s E(\hat{s}, \hat{r}) = (H_{\hat{s}}\hat{r})^T \dot{H}_{\hat{s}}\hat{r} - J^T \dot{H}_{\hat{s}}\hat{r} = 0 \qquad (\text{C.19})$$

and

$$\frac{1}{2}\nabla_r E(\hat{s}, \hat{r}) = H_{\hat{s}}^* H_{\hat{s}}\hat{r} - H_{\hat{s}}^* J = 0 \qquad (\text{C.20})$$

leads to

$$H_{\hat{s}}^* H_{\hat{s}}\hat{r} = H_{\hat{s}}^* J. \qquad (\text{C.21})$$

Now, the last equation is what defines the pseudo-inverse $H_{\hat{s}}^{\dagger}$ (see equation (2.17)), and therefore it is satisfied, by construction, when

$$\hat{r} = H_{\hat{s}}^{\dagger}J = H_{\hat{s}}^{\dagger}. \qquad (\text{C.22})$$

This shows that if $\hat{s}$ is a stationary point of E, its corresponding $\hat{r}$ must be of the form $H_{\hat{s}}^{\dagger}$. The computation of equation (C.13) returns

$$\frac{1}{2}\nabla_s \psi(\tilde{s}) = J^T H_{\tilde{s}}^{\perp} \dot{H}_{\tilde{s}}^{\perp} J = 0. \qquad (\text{C.23})$$

$\Longleftarrow$. Let us now assume that $\nabla_s \psi(\tilde{s}) = 0$, and let $\tilde{r} = H_{\tilde{s}}^{\dagger}$. We want to show that $\nabla_s E(\tilde{r}, \tilde{s}) = 0$; that is, equation (C.19) is satisfied with $\hat{s} = \tilde{s}$ (that equation (C.21) is satisfied follows directly from our choice of $\tilde{r}$ from equation (C.22)). To this end, notice that

$$\begin{aligned}H_s^{\perp}\left(H_s^{\dagger}\right)^* &= \left(H_s^{\dagger}\right)^* - H_s H_s^{\dagger}\left(H_s^{\dagger}\right)^*\\ &= \left(H_s^{\dagger}\right)^* - \left(H_s H_s^{\dagger}\right)^*\left(H_s^{\dagger}\right)^*\\ &= 0\end{aligned} \qquad (\text{C.24})$$

and therefore, substituting the expression of $\dot{H}_s^{\perp}$ (obtained in Lemma C.1) and the expression for $\tilde{r} = H_{\tilde{s}}^{\dagger}$ in equation (C.23) we obtain

$$\begin{aligned}0 &= \tfrac{1}{2}\nabla_s \psi(\tilde{s}) = J^T H_{\tilde{s}}^{\perp}\dot{H}_{\tilde{s}}^{\perp}J = -J^T H_{\tilde{s}}^{\perp}\dot{H}_{\tilde{s}}H_{\tilde{s}}^{\dagger}J\\ &= (H_{\tilde{s}}H_{\tilde{s}}^{\dagger}H_{\tilde{s}}\tilde{r})^T \dot{H}_{\tilde{s}}\tilde{r} - J^T \dot{H}_{\tilde{s}}\tilde{r}\\ &= \tfrac{1}{2}\nabla_s E(\tilde{r}, \tilde{s}).\end{aligned} \qquad (\text{C.25})$$

$\Longrightarrow$. Now, let equation (C.19) and equation (C.21) hold for $\hat{s}, \hat{r}$. All we need to show is that $\nabla_s \psi(\hat{s}) = 0$. In as much as $\hat{r}$ satisfies equation (C.22) because of equation (C.21), we can read equation (C.25) backward and have

$$
\begin{aligned}
0 &= \tfrac{1}{2} \nabla_s E(\hat{r}, \hat{s}) = (H_{\hat{s}} \hat{r})^T \dot{H}_{\hat{s}} \hat{r} - J^T \dot{H}_{\hat{s}} \hat{r} \\
&= \left(H_{\hat{s}} H_{\hat{s}}^{\dagger} J \right)^T \dot{H}_{\hat{s}} H_{\hat{s}}^{\dagger} J - J^T \dot{H}_{\hat{s}} H_{\hat{s}}^{\dagger} J \\
&= -J^T H_{\hat{s}}^{\perp} \dot{H}_{\hat{s}} H_{\hat{s}}^{\dagger} J \\
&= J^T H_{\hat{s}}^{\perp} \dot{H}_{\hat{s}}^{\perp} J \\
&= \tfrac{1}{2} \nabla_s \psi(\hat{s}),
\end{aligned}
\tag{C.26}
$$

which concludes the proof. $\qquad\qquad\qquad\qquad\qquad\qquad\qquad\qquad$ $\square$

C.4   Proof of Proposition 5.1

Proposition 5.1. *Let r_0 be a nonnegative real-valued function, and let the sequence r_k be defined according to equation (5.19). Then $E(s, r_{k+1}) \leq E(s, r_k)\ \forall\ k > 0$ and for all admissible surfaces s. Furthermore, equality holds if and only if $r_{k+1} = r_k$.*

Proof. The proof follows [Snyder et al., 1992]. From the definition of E in equation (5.2) we get

$$
E(\tilde{s}, r_{k+1}) - E(\tilde{s}, r_k) = -\int_{\Omega} J(y) \log \frac{I_{\tilde{s}}^{r_{k+1}}(y)}{I_{\tilde{s}}^{r_k}(y)} dy + \int_{\in \Omega} I_{\tilde{s}}^{r_{k+1}}(y) - I_{\tilde{s}}^{r_k}(y) dy.
\tag{C.27}
$$

The second integral in the above expression is given by

$$
\begin{aligned}
&\int_{\Omega} \int h_{\tilde{s}}(y, x) r_{k+1}(x) dx dy - \int_{\Omega} \int h_{\tilde{s}}(y, x) r_k(x) dx dy = \\
&= \int h_{\tilde{s},0}(x) r_{k+1}(x) dx - \int h_{\tilde{s},0}(x) r_k(x) dx,
\end{aligned}
$$

where we have defined $h_{\tilde{s},0}(x) = \int_{\Omega} h_{\tilde{s}}(y, x) dy$, and from the expression of r_{k+1} in equation (5.19) we have that the ratio in the first integral is

$$
\frac{I_{\tilde{s}}^{r_{k+1}}(y)}{I_{\tilde{s}}^{r_k}(y)} = \int F_{r_k}^{\tilde{s}}(x) \frac{h_{\tilde{s}}(y, x) r_k(x)}{I_{\tilde{s}}^{r_k}(y)} dx.
\tag{C.28}
$$

We next note that, from Jensen's inequality [Cover and Thomas, 1991],

$$
\log \left(\int F_{r_k}^{\tilde{s}}(x) \frac{h_{\tilde{s}}(y, x) r_k(x)}{I_{\tilde{s}}^{r_k}(y)} dx \right) \geq \int \frac{h_{\tilde{s}}(y, x) r_k(x)}{I_{\tilde{s}}^{r_k}(y)} \log \left(F_{r_k}^{\tilde{s}}(x) \right) dx
\tag{C.29}
$$

because the ratio $h_{\tilde{s}}(y,x)r_k(x)/I_{\tilde{s}}^{r_k}(y)$ integrates to 1 and is always positive, and therefore the expression in equation (C.27) is

$$
\begin{aligned}
E(\tilde{s}, r_{k+1}) - E(\tilde{s}, r_k) \;\leq\; & -\int_\Omega J(y) \int \log(F_{r_k}^{\tilde{s}}(x)) \frac{h_{\tilde{s}}(y,x)r_k(x)}{I_{\tilde{s}}^{r_k}(y)} dx\, dy \\
& + \int h_{\tilde{s},0}(x)r_{k+1}(x)dx - \int h_{\tilde{s},0}(x)r_k(x)dx \\
= \; & -\Phi\left(h_{\tilde{s},0}(\cdot)r_{k+1}(\cdot)\|h_{\tilde{s},0}(\cdot)r_k(\cdot)\right) \leq 0.
\end{aligned}
$$
(C.30)

Notice that the right-hand side of the last expression is still the I-divergence of two positive functions. Therefore, we have

$$
E(\tilde{s}, r_{k+1}) - E(\tilde{s}, r_k) \leq 0.
$$
(C.31)

Furthermore, Jensen's inequality becomes an equality if and only if $F_{r_k}^{\tilde{s}}$ is a constant; because the only admissible constant value is 1, because of the normalization constraint, we have $r_{k+1} = r_k$, which concludes the proof. $\quad\square$

Finally, we can conclude that the algorithm proposed generates a monotonically decreasing sequence of values of the cost function E.

Corollary C.2. *Let s_0, r_0 be admissible initial conditions for the sequences s_k and r_k defined from equation (5.19). The sequence $E(s_k, r_k)$ converges to a limit E^*:*

$$
\lim_{k \to \infty} E(s_k, r_k) = E^*.
$$
(C.32)

Proof. Follows directly from equation (5.5) and Proposition 5.1, together with the fact that the I-divergence is bounded from below by zero. $\quad\square$

C.5 Proof of Proposition 7.1

For ease of reading, we recall the statement of Proposition (7.1) of Section 7.2.

Theorem C.3. *Let $r \in L^2(\Omega)$ and $s \in H^1(\Omega)$ satisfy (7.10). Then, there exists a unique weak solution $u \in C(0, t_1; L^2(\Omega))$ of (7.9), satisfying*

$$
\int_0^{t_1} \int_\Omega \lambda(y)|\nabla u(y,t)|^2 \, dy\, dt \leq \int_\Omega r(y)^2 \, dy,
$$
(C.33)

where $\lambda(y) \geq 0$ denotes the minimal eigenvalue of $D(y)$.

Proof. We start by investigating an approximation to equation (7.9) with $D(y)$ replaced by $D^\epsilon(y) = D(y) + \epsilon I_d$ for $\epsilon > 0$. In this case, (7.9) is a parabolic problem with nondegenerate diffusion tensor, and standard theory (cf. [Evans, 1992]) shows that there exists a unique solution $u^\epsilon \in C(0, t_1; L^2(\Omega)) \cap L^2(0, t_1; H^1(\Omega))$. Moreover, one can deduce the a priori estimate

$$
\int_\Omega u^\epsilon(y,t)^2 \, dy + \int_0^\tau \int_\Omega (\epsilon + \lambda(y))|\nabla u^\epsilon(y,t)|^2 \, dy\, dt \leq \int_\Omega r(y)^2 \, dy
$$

for all $\tau \in (0, t_1]$. This estimate implies the existence of a weakly convergent subsequence u^{ϵ_k} as $\epsilon_k \to 0$ and by standard methods one can show that the limit $\tilde{u}$ is a weak solution satisfying (C.33). $\square$

Appendix D
Calibration of defocused images

Accommodation cues are the result of changing the geometry of the imaging device. For instance, in shape from defocus, one changes the lens settings or the position of the image plane, yielding a blurred image, where blur is related to the shape and radiance of the scene, which are the unknowns of interest.

Unfortunately, however, changing the geometry of the imaging device causes other artifacts that affect the image, but have little to do with the geometry or reflectance of the scene. These are "nuisance effects" that we need to avoid, either by processing the data to remove them, which we discuss in Section D.1, or by means of an imaging device that is, by design, unaffected by (or, more technically, invariant to) such nuisance. These are discussed in Section D.2.

Applying the techniques described in this book to images obtained with a photographic camera by changing the focus of the aperture would fail consistently, because of such artifacts. Fortunately, these are very simple to handle, as we show next.

D.1 Zooming and registration artifacts

Changing the focus on a commercial camera is usually accomplished by changing the relative position between the lens and the sensor (the image plane). In addition to causing blur, this also causes zooming, that is, a magnification of the entire image. Recalling the basic image formation model from Chapter 2, which we recall here in simplified form

$$y = v\frac{x}{s(x)},$$ (D.1)

one can easily see that the pixel coordinate y depends on the focus setting v, the coordinates x, and the depth map $s(x)$ at that point. Changing the focus setting v causes a rescaling of the coordinates y. Although at first it may seem that the magnification depends on the shape of the scene $s(x)$, this is not the case. Consider a pixel in position y_2 in an image I_2, obtained by changing the focus relative to another image I_1, where the corresponding pixel has coordinate y_1. From the previous equation we have that

$$\frac{y_1}{v_1} = \frac{x}{s(x)} = \frac{y_2}{v_2}$$ (D.2)

regardless of the shape $s(x)$. This means that we can seek the magnification factor between the two images

$$y_2 = \frac{v_2}{v_1}y_1$$ (D.3)

by minimizing their discrepancy, which yields a point-to-point correspondence between the two. The same holds for a number K of images obtained with different focus settings.

It may seem attractive, therefore, to try to establish correspondence between points in the two images, and then use the equation above to determine the magnification factor. There is a glitch: in computing the discrepancy between the two images we have to take into account the fact that they are blurred in different ways, so establishing point correspondence may be challenging. This can be alleviated by taking images of simple objects, say white circles in a dark background, so that certain statistics (say the centers) are easy to detect, and are relatively unaffected by blur. Such a simple scene is often called a *calibration target*, or rig.

An alternative approach is to make the magnification factor part of the unknowns and estimate it as part of the shape and reflectance inference problem. The reader can easily extend this idea to the algorithms presented throughout the book. In this section, we outline a simple procedure to calibrate an image using a simple calibration target.

Indeed, because depending on the camera defocusing can be achieved by a variety of mechanisms, including sliding the sensor, rotating the lens and so on, it is possible that artifacts more complex than zooming may affect the image. A simple model that we have found empirically to be general enough to enable removing nuisance artifacts and at the same time simple enough to allow a simple calibration procedures is the affine model, where

$$y_2 = Ay_1 + B,$$ (D.4)

where $A \in \mathbb{R}^{2 \times 2}$ is a nonsingular matrix that includes the magnification factor along its diagonal (it can be different for the x and y coordinates), and $B \in \mathbb{R}^2$ is an offset that takes into account the change in the center of the image, for instance, when the optical axis is not well aligned along the normal to the image plane.

There are many algorithms that one can employ to find the parameters A and B from two images, for instance, solving the following least-squares problem

$$\hat{A}, \hat{B} = \arg\min_{A,B} \|I_2(Ay_1 + B) - I_1(y_1)\|. \tag{D.5}$$

Clearly, whether this problem can be solved, and how well, depends on I_1 and I_2. Trivially, if they are both constants, then the problem is solved by any A and B.

It is therefore in our interest to design a calibration rig $I_1(\cdot)$ that makes the estimation of A and B well-posed. We have found that a dark planar rig with four white circles at the corners serves this purpose well. It is straightforward to fit ellipses to the circles, find their centers, $y_1^i, i = 1, \ldots, 4$, and then solve the linear least-squares problem

$$\hat{A}, \hat{B} = \arg\min_{A,B} \sum_{i=1}^{4} \|y_2^i - Ay_1^i - B\|. \tag{D.6}$$

In fact, we can further simplify the problem by assuming that A is diagonal, because we do not expect focus artifacts to include skewing the images. Then if we let $A = \text{diag}(a_1, a_2)$ and $B = [b_1 \; b_2]^T$, we have that

$$\hat{a}_1, \hat{a}_2, \hat{b}_1, \hat{b}_2 = \arg\min_{a_1,a_2,b_1,b_2} \left\| \bar{Y} \begin{bmatrix} a_1 \\ a_2 \\ b_1 \\ b_2 \end{bmatrix} - Y_0 \right\|, \tag{D.7}$$

where we have defined

$$\bar{Y} \doteq \begin{bmatrix} y_{11}^1 & 0 & 1 & 0 \\ 0 & y_{12}^1 & 0 & 1 \\ \vdots & \vdots & \vdots & \vdots \\ y_{11}^4 & 0 & 1 & 0 \\ 0 & y_{12}^4 & 0 & 1 \end{bmatrix} \tag{D.8}$$

and

$$P \doteq \begin{bmatrix} y_{21}^1 \\ y_{22}^1 \\ \vdots \\ y_{21}^4 \\ y_{22}^4 \end{bmatrix} \tag{D.9}$$

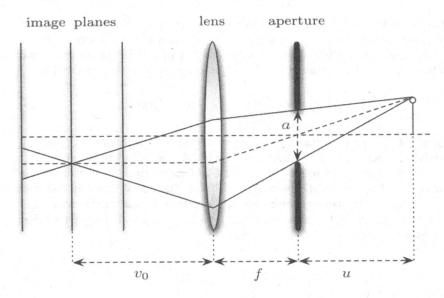

Figure D.1. Telecentric optics. An additional aperture is placed in front of the lens so that for any focus setting, the COC has always the same center.

This problem is solved easily in the least-squares sense, via the pseudo-inverse, introduced for the more general case in Section 2.1.4:

$$
\begin{bmatrix} \hat{a}_1 \\ \hat{a}_2 \\ \hat{b}_1 \\ \hat{b}_2 \end{bmatrix} = \bar{Y}^{\dagger} Y_0.
\tag{D.10}
$$

An alternative to preprocess the images is to devise an image formation apparatus that minimizes the magnification effects due to changes in focus settings.

D.2   Telecentric optics

A telecentric optical system is one designed so that optical rays are parallel to the optical axis. In idealized mathematical terms, a telecentric optical system is modeled by a parallel (orthographic) projection, rather than a central (perspective) projection that we discussed in Chapter 2 (see Figure D.1).

Today there are commercial optical systems that implement telecentric optics; they are typically multilens systems designed to approximate parallel projection, and are usually designed for a particular application. In the particular application to shape from defocus, or more in general focus analysis, [Watanabe and Nayar, 1997] have described a telecentric optical system that can be implemented in the laboratory. In this book, we concentrate on the algorithmic aspects of shape from defocus, and therefore refer the reader to [Watanabe and Nayar, 1997] or to the commercial catalogues of major optical design companies.

Appendix E

MATLAB® implementation of some algorithms

E.1 Least-squares solution (Chapter 4)

● Least Squares Main Script ●

```
%
% Geometric Blind Deconvolution script
%

clear all
close all

Yes = 1;
No = 0;
fprintf('Shape from Defocus (Least Squares)\n');
notValid = Yes;
while notValid
    method = input(['Which method do you want to test?'...
        '\nChoose generate, learn, overlap: '],'s');
    notValid = ~strcmp(method,'generate')&...
        ~strcmp(method,'learn')&...
        ~strcmp(method,'overlap');
end
%%%%%%%%%%%%%%%%%%%%%%
% compute operators %
%%%%%%%%%%%%%%%%%%%%%%
```

```
% ni is the square patch size
% the larger ni the slower the algorithm
notValid = Yes;
while notValid
    ni = input(['Choose the patch size '...
        '\n3(fast&noisy),5,7,9(slow&smooth): '],'s');
    notValid = ~(ni=='3')&...
        ~(ni=='5')&...
        ~(ni=='7')&...
        ~(ni=='9');
end
ni = str2num(ni);
% fix rank of operators
ranks = ni*ni; % operator ranks
if strcmp(method,'generate')
    % compute operators when PSF is known
    Hp = generate_operators(ni,ranks);
elseif strcmp(method,'learn')
    % compute operators when PSF is unknown
    % learning approach; non overlapping patches
    Hp = learn_operators(ni,ranks);
else
    % compute operators when PSF is unknown
    % learning approach; overlapping patches
    Hp = learn_operators_overlap(ni,ranks);
end
%%%%%%%%%%%%%%
% load images %
%%%%%%%%%%%%%%
load DataSet

%%%%%%%%%%%%%%%%%%
% Estimate Depth %
%%%%%%%%%%%%%%%%%%
Depth = estimate_depth(I1,I2,Hp);

% optional smoothing
FilteredDepth = mediansmoothing(Depth);

return
```

●generate_operators●

```
function Hp = generate_operators(ni,ranks)
% Generates a family of orthogonal operators
% for patches of size ni x ni.
% The operators are derived by assuming Gaussian
% point-spread functions.

%
% ni = square patch size
```

```
%
auto_rank = 0;
if nargin<2
   % if no rank is provided, determine it from PSF
   auto_rank = 1;
end
if nargin<1
   error('Too few arguments!');
end
%%%%%%%%%%%%%%%%%%%%%%%
% image coordinates %
%%%%%%%%%%%%%%%%%%%%%%%
[X,Y] = meshgrid([-(ni-1)/2:(ni-1)/2],[-(ni-1)/2:(ni-1)/2]);
x1 = repmat(X(:),1,ni*ni);
x2 = repmat(Y(:),1,ni*ni);
Dx = x1-x1';
Dy = x2-x2';
tolerance = 1e-5;
%%%%%%%%%%%%%%%%%%%%%%%%%%%%%%%%%
% parameters of the optics %
%%%%%%%%%%%%%%%%%%%%%%%%%%%%%%%%%
F = 35e-3; % focal length
Fnumb = 4; % F-number
D = F/Fnumb; % aperture (actual lens diameter)
gamma = .8e4; %calibration parameter (CCD pixel size)
z0 = .53; % distance of near focal plane from camera
z1 = .85; % distance of far focal plane from camera
p = 1./(1/F-1./[z0 z1]);% distance of lens from CCD (two images)
%%%%%%%%%%%%%%%%%%%%%%%%%%%%%%%%
% synthetic depth map %
%%%%%%%%%%%%%%%%%%%%%%%%%%%%%%%%
numlevels = 50; % number of operators (i.e., number of depth
                %          levels we distinguish
% positions of equifocal planes
depthlevel = [z0:(z1-z0)/(numlevels+1):z1];
%%%%%%%%%%%%%%%%%%%%%%%%%%%%%%%%%%%%%%%%%
% generate orthogonal operators %
%%%%%%%%%%%%%%%%%%%%%%%%%%%%%%%%%%%%%%%%%
fprintf('Computing orthogonal operators (Gaussian kernel)\n');
fprintf('%i Levels\n    ',numlevels)
for k=1:numlevels
   fprintf('\b\b\b\b[%.2i]',k);
   Depth = depthlevel(k);
   for j1=1:length(p)
      % blurring radius
      sigma1 = (gamma*D/2*abs(1-p(j1)*(1/F-1/Depth))).^2;
      % integrate over pixel area
      if (sigma1>tolerance)
         wsi = Dx/sqrt(2)/sigma1;
         wsj = Dy/sqrt(2)/sigma1;
         wdx = .5/sqrt(2)/sigma1;
         derfx = erf(wsi+wdx)-erf(wsi-wdx);
         derfy = erf(wsj+wdx)-erf(wsj-wdx);
```

```
            PSF1 = .25*derfx.*derfy;
        else
            PSF1 = (Dx==0).*(Dy==0); % Dirac delta
        end
        % compute a defocused image for each focus setting
        for j2=1:length(p)
            % blurring radius
            sigma2 = (gamma*D/2*abs(1-p(j2)*(1/F-1/Depth))).^2;
            % integrate over pixel area
            if (sigma2>tolerance)
                wsi = Dx/sqrt(2)/sigma2;
                wsj = Dy/sqrt(2)/sigma2;
                wdx = .5/sqrt(2)/sigma2;
                derfx = erf(wsi+wdx)-erf(wsi-wdx);
                derfy = erf(wsj+wdx)-erf(wsj-wdx);
                PSF2 = .25*derfx.*derfy;
            else
                PSF2 = (Dx==0).*(Dy==0); % Dirac delta
            end
            M((j1-1)*ni*ni+[1:ni*ni],(j2-1)*ni*ni+[1:ni*ni]) = ...
                PSF1*PSF2';
        end
    end
    HHT(:,:,k) = M;
end
fprintf('Done.\n');
Hp = zeros(ni*ni*2,ni*ni*2,numlevels,length(ranks));
% extract regularized orthogonal operator
for focus=1:numlevels
    [u,s,v] = svd(HHT(:,:,focus));
    if auto_rank
        ranks = rank(s);
    end
    for rnk=1:length(ranks)
        Hp(:,:,focus,rnk) = ...
            u(:,ranks(rnk):size(u,2))*u(:,ranks(rnk):size(u,2))';
    end
end
% save all data
save Operators Hp
return
```

•learn_operators•

```
function Hp = learn_operators(ni,ranks,training_seq)
% Learns a family of orthogonal operators
% for patches of size ni x ni from defocused images.

%
% ni = square patch size
%
```

```
if nargin<3
    need_training = 1;
    training_seq = zeros(ni*ni*2,ni*ni*4);
else
    need_training = 0;
    error('Not yet implemented');
end
auto_rank = 0;
if nargin<2
    % if no rank is defined, then compute rank of training set
    auto_rank = 1;
end
if nargin<1
    error('Too few arguments!');
end
%%%%%%%%%%%%%%%%%%%%%%%
% image coordinates %
%%%%%%%%%%%%%%%%%%%%%%%
X = ones(ni,1)*[-(ni-1)/2:(ni-1)/2];
Y = [-(ni-1)/2:(ni-1)/2]'*ones(1,ni);
%%%%%%%%%%%%%%%%%%%%%%%
% optics parameters %
%%%%%%%%%%%%%%%%%%%%%%%
F = 35e-3; % focal length
Fnumb = 4; % F-number
D = F/Fnumb; % aperture (effective lens diameter)
gamma = .8e4; %calibration parameter (CCD pixel size)
z0 = .53; % distance of near focal plane from camera
z1 = .85; % distance of far focal plane from camera
p = 1./(1/F-1./[z0 z1]);% distance of lens from CCD (two images)
%%%%%%%%%%%%%%%%%%%%%%%%%%%
% synthetic depth map %
%%%%%%%%%%%%%%%%%%%%%%%%%%%
numlevels = 50; % number of operators (i.e., number of depth
                %    levels we distinguish
% positions of equifocal planes
depthlevel = [z0:(z1-z0)/(numlevels+1):z1];
%%%%%%%%%%%%%%%%%%%%%%%%%%%%%%%%%%%
% generate defocused images %
%%%%%%%%%%%%%%%%%%%%%%%%%%%%%%%%%%%
randn('state',0);
Nit = 5;
pad = 3;
if need_training
    fprintf('Computing orthogonal operators (learning)\n');
    fprintf('%i Levels\n    ',numlevels)
    for j=1:numlevels
        fprintf('\b\b\b\b[%.2i]',j);
        % generate training set (Gaussian PSF)
        TS = [];
        DepthMap = depthlevel(j);
        for i=1:size(training_seq,2)
            Rad = abs(randn(ni+2*pad))*200+100;
```

```
        J = zeros(ni+2*pad,ni+2*pad,length(p));
        % use heat equation to compute convolution (Gaussian PSF)
        for k=1:length(p)
            Mask = (gamma*D/2*abs(1-p(k).*(1/F-1/DepthMap))).^2;
            J(:,:,k) = del2(defocusUnif(Rad,Mask));
        end
        J = J(pad+[1:ni],pad+[1:ni],:);
        TS = [TS J(:)];
    end
    [u,s,v] = svd(TS);
    if auto_rank
        ranks = rank(s);
    end
    Hp(:,:,j) = u(:,2*ni*ni-ranks:ni*ni*2)*...
        u(:,2*ni*ni-ranks:ni*ni*2)';
  end
end
fprintf('Done.\n');
% save all data
save Operators Hp
return
```

•learn_operators_overlap•

```
function Hp = learn_operators_overlap(ni,ranks,training_seq)
% Learns a family of orthogonal operators
% for patches of size ni x ni from defocused images.

%
% ni = square patch size
%
if nargin<3
   need_training = 1;
   training_seq = zeros(ni*ni*2,ni*ni*4);
else
   need_training = 0;
   error('Not yet implemented');
end
auto_rank = 0;
if nargin<2
   % if no rank is defined, then compute rank of training set
   auto_rank = 1;
end
if nargin<1
   error('Too few arguments!');
end
%%%%%%%%%%%%%%%%%%%%%%
% image coordinates %
%%%%%%%%%%%%%%%%%%%%%%%
m = ni*3;
n = ni*ni*5;
```

```
X = ones(m,1)*[-(n-1)/2:(n-1)/2];
Y = [-(m-1)/2:(m-1)/2]'*ones(1,n);
%%%%%%%%%%%%%%%%%%%%%
% optics parameters %
%%%%%%%%%%%%%%%%%%%%%
F = 35e-3; % focal length
Fnumb = 4; % F-number
D = F/Fnumb; % aperture (effective lens diameter)
gamma = .8e4; %calibration parameter (CCD pixel size)
z0 = .53; % distance of near focal plane from camera
z1 = .85; % distance of far focal plane from camera
p = 1./(1/F-1./[z0 z1]);% distance of lens from CCD (two images)
%%%%%%%%%%%%%%%%%%%%%%%%%%%
% synthetic depth map %
%%%%%%%%%%%%%%%%%%%%%%%%%%%
numlevels = 50; % number of operators (i.e., number of depth
                % levels we distinguish
% positions of equifocal planes
depthlevel = [z0:(z1-z0)/(numlevels+1):z1];
%%%%%%%%%%%%%%%%%%%%%%%%%%%%%%%%%%%
% generate defocused images %
%%%%%%%%%%%%%%%%%%%%%%%%%%%%%%%%%%%
randn('state',0);
Nit = 5;
if need_training
    fprintf('Computing orthogonal operators (learning)\n');
    fprintf('%i Levels\n     ',numlevels)
    for j=1:numlevels
        fprintf('\b\b\b\b[%.2i]',j);
        % generate training set (Gaussian PSF)
        DepthMap = depthlevel(j);
        Rad = abs(randn(m,n))*200+100;
        % use heat equation to compute convolution (Gaussian PSF)
        J = zeros(m,n,length(p));
        for k=1:length(p)
            Mask = (gamma*D/2*abs(1-p(k).*(1/F-1/DepthMap))).^2;
            J(:,:,k) = del2(defocusUnif(Rad,Mask));
        end
        TS = zeros(ni*ni*2,size(J,2)-2*ni);
        for i=1:size(J,2)-2*ni
            TS(:,i) = reshape(J(ni+[1:ni],ni+[1:ni]+(i-1),:),...
                ni*ni*length(p),1);
        end
        [u,s,v] = svd(TS);
        if auto_rank
            ranks = rank(s);
        end
        Hp(:,:,j) = u(:,2*ni*ni-ranks:ni*ni*2)*...
            u(:,2*ni*ni-ranks:ni*ni*2)';
    end
end
fprintf('Done.\n');
% save all data
```

```
save Operators Hp
return
```

•defocusUnif•

```
function [I] = defocusUnif(Radiance,sigma)
% Generate defocused images (uniform blur)

sigmamax = max(sigma(:));
win = ceil(sigmamax*3);

% extend domain
Radiance_ext = ones(size(Radiance,1)+...
    2*win,size(Radiance,2)+...
    2*win,size(Radiance,3));
Radiance_ext(win+1:end-win,win+1:end-win,:) = Radiance;
if win>0
    Radiance_ext(1:win,:,:) = [...
        Radiance(win+1:-1:2,win+1:-1:2,:) ...
        Radiance(win+1:-1:2,:,:) ...
        Radiance(win+1:-1:2,end-1:-1:end-win,:)];
    Radiance_ext(end-win+1:end,:,:) = [...
        Radiance(end-1:-1:end-win,end-1:-1:end-win,:) ...
        Radiance(end-1:-1:end-win,:,:) ...
        Radiance(end-1:-1:end-win,end-1:-1:end-win,:)];
    Radiance_ext(win+1:end-win,1:win,:) = ...
        Radiance(:,win+1:-1:2,:);
    Radiance_ext(win+1:end-win,end-win+1:end,:) = ...
        Radiance(:,end-1:-1:end-win,:);
end

[M,N,K] = size(Radiance_ext);
K = length(sigma); % number of focus settings
tolerance = 1e-5; % threshold for Dirac delta approx. of kernel

W = win*2+1; % max size of the kernel window
% compute window coordinates
[wi,wj] = meshgrid([-win:win],[-win:win]);
% allocate space for the defocused images
I = ones(M,N,K);

% compute a defocused image for each focus setting
for k=1:K
    % integrate over pixel area
    if (sigma(k)>tolerance)
        wsi = wi/sqrt(2)/sigma(k);
        wsj = wj/sqrt(2)/sigma(k);
        wdx = .5/sqrt(2)/sigma(k);
        derf = erf(wsi+wdx)-erf(wsi-wdx);
        PSF = .25*derf.*derf';
    else
```

```
        PSF = (wi==0).*(wj==0); % Dirac delta
    end
    I(:,:,k) = conv2(Radiance_ext(:,:,k),PSF,'same');
end
% restore original domain
I = I(win+1:M-win,win+1:N-win,:);

return
```

●estimate_depth●

```
function Depth = estimate_depth(I1,I2,Hp)
% This function returns a depth estimate from
% images I1 (near focused) and I2 (far focused).
% Depth estimation is performed by extracting
% the orthogonal operators family from HHT, which can
% be passed as an argument, or will be retrieved from
% the file Operators.mat

if nargin<3
    fid=fopen('Operators.mat','r');
    fclose(fid);
    if fid==-1
        error('No operators found. Run generate_operators first.')
    end
    load Operators
end
if nargin<2
    error('Too few arguments!');
end
numlevels = size(Hp,3);
ni = sqrt(size(Hp,1)/2);
if ni~=floor(ni)
    error('Orthogonal operators have incorrect size!');
end
%%%%%%%%%%%%%%%%%%%%%%%%%%%%%%%%%%
% image linear prefiltering %
%%%%%%%%%%%%%%%%%%%%%%%%%%%%%%%%%%
% eliminate brightness component
dI1 = del2(I1);
dI2 = del2(I2);
[mI,nI] = size(I1);
% compute depth level
fprintf('Computing depth levels\n');
fprintf('%i Iterations\nIteration      ',numlevels)
Cost = zeros(mI*nI,numlevels);
C = zeros(mI*nI,size(Hp,1));
for i = 1:numlevels
    fprintf('\b\b\b\b\b[%3i]',i);
    for k=1:size(Hp,1)
        P1 = reshape(Hp(k,ni*ni:-1:1,i,1),ni,ni);
```

```
        P2 = reshape(Hp(k,ni*ni+[ni*ni:-1:1],i,1),ni,ni);
        C(:,k) = reshape(conv2(dI1,P1,'same')+...
            conv2(dI2,P2,'same'),mI*nI,1);
    end
    Flt = C*Hp(:,:,i);
    Cost(:,i) = sum(Flt.^2,2);
end
% find minimum of the cost functional
[dummy,Depth] = min(Cost,[],2);
Depth = reshape(Depth,mI,nI);
fprintf('\ndone.\n');
return
```

●mediansmoothing●

```
function F = mediansmoothing(f)
% This function performs median filtering and then
% Gaussian smoothing

% window of median filtering and Gaussian smoothing
win = 7;

% median filtering
k = 0;
F = zeros(win*win,(size(f,1)-win+1)*(size(f,2)-win+1));
for i=0:win-1
    for j=0:win-1
        k = k+1;
        F(k,:) = reshape(f(i+[1:size(f,1)-win+1],...
                j+[1:size(f,2)-win+1]),...
                1,(size(f,1)-win+1)*(size(f,2)-win+1));
    end
end
F = median(F,1);
Temp = reshape(F,size(f,1)-win+1,size(f,2)-win+1);
F = median(f(:))*ones(size(f));
F(floor((win-1)/2)+[1:size(f,1)-win+1],...
   floor((win-1)/2)+[1:size(f,2)-win+1]) = Temp;

% smoothing
[x,y] = meshgrid([-(win-1)/2:(win-1)/2]/(win-1)*5,...
                [-(win-1)/2:(win-1)/2]/(win-1)*5);
weights = exp(-x.^2-y.^2);
weights = weights/sum(weights(:));
F = conv2(F-mean(F(:)),weights,'same')+mean(F(:));
return
```

E.2   I-divergence solution (Chapter 5)

•I-Divergence Main Script•

```
% Depth from Defocus and Image Restoration
% by Minimizing I-Divergence

clear all
close all

No = 0;
Yes = 1;

% set the maximum number of iterations
MaxIterations = 100;

fprintf('Image restoration and shape estimation\n');
fprintf('I-divergence approach\n\n');
% shape type (wave, slope, box, sin)
notValid = Yes;
while notValid
    shape = input(['What test do you want to perform?'...
        '\nChoose wave, slope, box, sin, plane, realdata: '],'s');
    notValid = ~strcmp(shape,'wave')&...
        ~strcmp(shape,'slope')&...
        ~strcmp(shape,'box')&...
        ~strcmp(shape,'sin')&...
        ~strcmp(shape,'plane')&...
        ~strcmp(shape,'realdata');
end

%%%%%%%%%%%%%%%%%%%%%%%%%%%%%%%%%%%%%%%%
% Data set initialization
%%%%%%%%%%%%%%%%%%%%%%%%%%%%%%%%%%%%%%%%
% choose between loading or generating defocused images
LoadImage = strcmp(shape,'realdata');
if LoadImage == No
    % generate synthetic image
    fprintf('Generating synthetic dataset...');
    % set image size
    M = 101; % use odd numbers
    N = 101; % use odd numbers
    % lens parameters
    F = 12e-3; % focal length
    Fnumb = 2; % F number
    D = F/Fnumb; % aperture (effective lens diameter)
    gamma = 2e4; % calibration parameter
    p = 1./(1/F-1./[.52 .85]); % distance CCD-lens
    K = length(p); % number of focus settings
    % generate depth map
```

```matlab
[X,Y] = meshgrid([-(N-1)/2:(N-1)/2],[-(M-1)/2:(M-1)/2]);
switch shape
    case 'wave'
        DepthMap = (1+cos(-(X.^2+Y.^2)/200))./...
            sqrt(X.^2/8600+Y.^2/8600+1)/2*.33+.52;
    case 'slope'
        % along X
        DepthMap = (X-min(X(:)))/(max(X(:))-min(X(:)))*.33+.52;
    case 'box'
        block1 = floor(M/5:M/5*3);
        block2 = floor(M/5*2:M/5*4);
        block3 = floor(N/10:N/2);
        block4 = floor(N/10*4:N/10*9);
        DepthMap = ones(M,N)*.52;
        DepthMap(block1,block3) = ...
            DepthMap(block1,block3)+.2;
        DepthMap(block2,block4) = ...
            DepthMap(block2,block4)+.13;
    case 'sin'
        DepthMap = (sin(X/3)+1)/2*.33+.52;
    otherwise % plane
        DepthMap = ones(M,N)*.33/2+.52;
end
% compute corresponding blur maps
for k=1:K
    BlurMap(:,:,k) = gamma*D/2*abs(1-p(k).*(1/F-1./DepthMap));
end
% generate random texture (with different intensities)
Radiance = [50*rand(floor(M/3),N);...
    120*rand(floor(M/3),N);...
    255*rand(M-2*floor(M/3),N)]+1; % skip zero
% add space-varying blurring to challenge algorithm with
% different textures (high-freq to low-freq)
TextureDepth = [zeros(M,floor(N/3))...
    .4*ones(M,floor(N/3))...
    .6*ones(M,N-2*floor(N/3))];
Radiance = defocusApprox(Radiance,TextureDepth);
% generate defocused images
I = defocusApprox(Radiance,BlurMap);
% select the domain where the data term is considered
win = 3;
mask = ones(M,N);
mask(1:win,:) = 0;
mask(end-win+1:end,:) = 0;
mask(:,1:win) = 0;
mask(:,end-win+1:end) = 0;
dt_data = 2e-1;% data term step
dt_reg = 2e-1; % regularization term step
maxDelta = 1; % max amount of relative blur
% regularization constant
% to avoid division by zero in the preconditioner
epsilon = max(I(:))/255;
fprintf('done\n');
```

```
else
    % load images
    fprintf('Loading dataset...');
    load('DataSet.mat');
    % first resize
    scalingFactor = 1/2;
    I1 = imresize(I1,scalingFactor); % limit number of computations
    I2 = imresize(I2,scalingFactor); % limit number of computations
    % select region of interest
    I(:,:,2) = I1(1:end-10,60:end-50);
    I(:,:,1) = I2(1:end-10,60:end-50);
    I = I+(I<1e-3);% make sure images are strictly positive
    [M,N,K] = size(I);
    % lens parameters
    F = 12e-3; % focal length
    Fnumb = 2; % F number
    D = F/Fnumb; % aperture (effective lens diameter)
    gamma = 3e4; % calibration parameter
    p = 1./(1/F-1./[.52 .85]); % distance CCD-lens
    % select the domain where the data term is considered
    win = 3;
    mask = ones(M,N);
    mask(1:win,:) = 0;
    mask(end-win+1:end,:) = 0;
    mask(:,1:win) = 0;
    mask(:,end-win+1:end) = 0;
    dt_data = 2e-1; % data term step
    dt_reg = 1e0; % regularization term step
    maxDelta = 1; % max amount of relative blur
    % regularization constant
    % to avoid division by zero in the preconditioner
    epsilon = 40*max(I(:))/255;
    K = length(p); % number of focus settings
    fprintf('done\n')
end

% discrete gradient step
dx = 1e-3;

% preconditioning threshold
thresh = 6e-2;

% choose whether to use the fast initialization or not
% init type (Y,N)
notValid = Yes;
while notValid
    init = input(['Do you want to use '...
                  'the fast initialization (Y/N)?'],'s');
    notValid = ~strcmp(init,'Y')&...
        ~strcmp(init,'N');
end

if strcmp(init,'Y')
```

```
    % initialize depth map and radiance
    [DepthMap_e,Radiance_e] = initialize(I,maxDelta,win,...
                              gamma,F,D,p,'I-Divergence');
else
    % initialize depth map with a plane
    DepthMap_e = sum(p)*F/(sum(p)-2*F)*ones(M,N);
    % initialize radiance with one image
    Radiance_e = I(:,:,1);
end

% start deblurring iterations
fprintf(['Image restoration and shape estimation'...
    ' (%3i iterations)\nIteration     '],MaxIterations);
for i=1:MaxIterations
    fprintf(['\b\b\b\b\b' ...
            '[%3i]'],i);
    % compute I-div gradient
    for k=1:K
        % compute corresponding blur maps
        BlurMap_e(:,:,k) = gamma*D/2*abs(1-p(k).*...
            (1/F-1./DepthMap_e));
        BlurMap_dx(:,:,k) = gamma*D/2*abs(1-p(k).*...
            (1/F-1./(DepthMap_e+dx)));
        % limit the amount of blur (to prevent out
        % of memory in defocusApprox)
        BlurMap_e(:,:,k) = BlurMap_e(:,:,k).*...
            (BlurMap_e(:,:,k)<2)+2*(BlurMap_e(:,:,k)>=2);
        BlurMap_dx(:,:,k) = BlurMap_dx(:,:,k).*...
            (BlurMap_dx(:,:,k)<2)+2*(BlurMap_dx(:,:,k)>=2);
    end
    % compute synthetic defocused images and gradients
    [I_e,F_e,I_dx] = defocusApprox(Radiance_e,BlurMap_e,I,...
                     BlurMap_dx);
    % update radiance (Lucy-Richardson iteration)
    Radiance_e = Radiance_e.*F_e;
    % compute gradient of image wrt depth map
    dIds = (I_dx-I_e)/dx;
    % compute gradient of idiv wrt depth map
    dEds = sum((1-I./I_e).*dIds,3);
    % compute preconditioning (positive definite) kernel
    precond = (epsilon+sum(abs(dIds),3)).^2./(epsilon+sum(I_e,3));
    dEds = dEds./precond;
    % precondition gradient
    dEds = (abs(dEds) <  thresh).*dEds+...
            (abs(dEds) >= thresh).*thresh.*sign(dEds);
    % at the boundary use only regularization
    dDatads = dEds.*mask;
    % combine data term and regularization term
    % make sure the iterations are an even number
    nIt = 2*(floor(dt_reg/1e-1)+1);
    beta_data = dt_data/nIt;
    beta_regularization = dt_reg/nIt;
    for tt=1:nIt
```

```
        % compute gradient of the cost functional
        % (data term + regularization) wrt the depth map
        if tt-floor(tt/2)*2==0
            dDepth = beta_data*dDatads - beta_regularization*...
                laplacian(DepthMap_e,'forward');
        else
            dDepth = beta_data*dDatads - beta_regularization*...
                laplacian(DepthMap_e,'backward');
        end
        % update depth map
        DepthMap_e = DepthMap_e - dDepth;
    end
end
fprintf('\ndone\n');

return
```

•laplacian•

```
function L = laplacian(A,dir)
% Compute the laplacian of A using either the forward
% difference scheme or the backward difference scheme

% padding size
win = 3;
if (size(A,3) >  1)          | ...
   (size(A,2) == 1)          | ...
   (size(A,1) == 1)          | ...
   (size(A,1) <  2*win+1)    | ...
   (size(A,2) <  2*win+1)
    error('function:laplacian:wrongInput',...
        ['Function laplacian can only handle matrices\n'...
         'and they must be at least %ix%i.'],2*win+1,2*win+1);
end
A = squeeze(A);

[M,N] = size(A);

% compute the regularization term
sx = zeros(M,N);
sy = zeros(M,N);
sxx = zeros(M,N);
syy = zeros(M,N);
% alternate the finite difference method
if strcmp(dir,'forward')
    sx(:,2:N) = diff(A,1,2);
    %sx(:,1) = sx(:,2);
    sy(2:M,:) = diff(A,1,1);
    %sy(1,:) = sy(2,:);
    sxx(:,1:N-1) = diff(sx,1,2);
    sxx(:,N) = sxx(:,N-1);
```

```
    syy(1:M-1,:) = diff(sy,1,1);
    syy(M,:) = syy(M-1,:);
else
    sx(:,1:N-1) = diff(A,1,2);
    %sx(:,N) = sx(:,N-1);
    sy(1:M-1,:) = diff(A,1,1);
    %sy(M,:) = sy(M-1,:);
    sxx(:,2:N) = diff(sx,1,2);
    sxx(:,1) = sxx(:,2);
    syy(2:M,:) = diff(sy,1,1);
    syy(1,:) = syy(2,:);
end
L = sxx+syy;

return
```

•defocusApprox•

```
function [I,F,I_dx] = defocusApprox(Radiance,BlurMap,I0,dBlurMap)
% Generate defocused images and deblurring gradient using
% approximate-but-fast kernel

if nargin<4
    sigmamax = max(BlurMap(:));
else
    sigmamax = max([BlurMap(:);dBlurMap(:)]);
end
win = ceil(sigmamax*3);

% extend domain
Radiance_ext = ones(size(Radiance)+2*win);
Radiance_ext(win+1:end-win,win+1:end-win) = Radiance;
Radiance_ext(1:win,:) = [Radiance(win+1:-1:2,win+1:-1:2) ...
    Radiance(win+1:-1:2,:) ...
    Radiance(win+1:-1:2,end-1:-1:end-win)];
Radiance_ext(end-win+1:end,:) = [...
    Radiance(end-1:-1:end-win,end-1:-1:end-win) ...
    Radiance(end-1:-1:end-win,:) ...
    Radiance(end-1:-1:end-win,end-1:-1:end-win)];
Radiance_ext(win+1:end-win,1:win) = Radiance(:,win+1:-1:2);
Radiance_ext(win+1:end-win,end-win+1:end) = ...
    Radiance(:,end-1:-1:end-win);

[M,N] = size(Radiance_ext);
K = size(BlurMap,3); % number of focus settings

BlurMap_ext = ones(M,N,K);
BlurMap_ext(win+1:end-win,win+1:end-win,:) = BlurMap;
BlurMap_ext(1:win,:,:) = [BlurMap(win+1:-1:2,win+1:-1:2,:) ...
    BlurMap(win+1:-1:2,:,:) ...
    BlurMap(win+1:-1:2,end-1:-1:end-win,:)];
```

```
BlurMap_ext(end-win+1:end,:,:) = [...
    BlurMap(end-1:-1:end-win,end-1:-1:end-win,:) ...
    BlurMap(end-1:-1:end-win,:,:) ...
    BlurMap(end-1:-1:end-win,end-1:-1:end-win,:)];
BlurMap_ext(win+1:end-win,1:win,:) = BlurMap(:,win+1:-1:2,:);
BlurMap_ext(win+1:end-win,end-win+1:end,:) = ...
    BlurMap(:,end-1:-1:end-win,:);

tolerance = 1e-5; % threshold for Dirac delta approx. of kernel

W = win*2+1; % max size of the kernel window
% compute window coordinates
[wi,wj] = meshgrid([-win:win],[-win:win]);
% precompute integrating grid
wip4 = (wi+.5)/sqrt(2)*1.694;
wim4 = (wi-.5)/sqrt(2)*1.694;
% compute pixel coordinates
[ti,tj] = meshgrid([win+1:N-win],[win+1:M-win]);
ti = ti(:);
tj = tj(:);
% allocate memory for the kernel
H = zeros((M-W+1)*(N-W+1),W*W);
% allocate space for the defocused images
I = ones(M,N,K);
% allocate space for the kernel
Kernel = zeros(M,N,W,W);
if nargin>2
    I0_ext = ones(M,N,K);
    I0_ext(win+1:end-win,win+1:end-win,:) = I0;
    I0_ext(1:win,:,:) = [I0(win+1:-1:2,win+1:-1:2,:) ...
        I0(win+1:-1:2,:,:) ...
        I0(win+1:-1:2,end-1:-1:end-win,:)];
    I0_ext(end-win+1:end,:,:) = [...
        I0(end-1:-1:end-win,end-1:-1:end-win,:) ...
        I0(end-1:-1:end-win,:,:) ...
        I0(end-1:-1:end-win,end-1:-1:end-win,:)];
    I0_ext(win+1:end-win,1:win,:) = I0(:,win+1:-1:2,:);
    I0_ext(win+1:end-win,end-win+1:end,:) = ...
        I0(:,end-1:-1:end-win,:);
    % initialize Lucy-Richardson variables
    H0 = zeros(M,N,K);
    F0 = zeros(M,N,K);
    F = ones(M,N);
    Ratio = zeros(M,N,K);
end
if nargin>3
    dBlurMap_ext = ones(M,N,K);
    dBlurMap_ext(win+1:end-win,win+1:end-win,:) = dBlurMap;
    dBlurMap_ext(1:win,:,:) = [...
        dBlurMap(win+1:-1:2,win+1:-1:2,:) ...
        dBlurMap(win+1:-1:2,:,:) ...
        dBlurMap(win+1:-1:2,end-1:-1:end-win,:)];
    dBlurMap_ext(end-win+1:end,:,:) = [...
```

```
            dBlurMap(end-1:-1:end-win,end-1:-1:end-win,:) ...
            dBlurMap(end-1:-1:end-win,:,:) ...
            dBlurMap(end-1:-1:end-win,end-1:-1:end-win,:)];
        dBlurMap_ext(win+1:end-win,1:win,:) = ...
            dBlurMap(:,win+1:-1:2,:);
        dBlurMap_ext(win+1:end-win,end-win+1:end,:) = ...
            dBlurMap(:,end-1:-1:end-win,:);
        % initialize gradient wrt shape
        I_dx = ones(M,N,K);
        H_dx = zeros((M-W+1)*(N-W+1),W*W);
end

transpose = reshape([1:W^2],W,W)';

i = ti(1:(M-W+1)*(N-W+1));
j = tj(1:(M-W+1)*(N-W+1));
ind = j+(i-1)*size(BlurMap_ext,1)-...
        size(BlurMap_ext,1)*size(BlurMap_ext,2);

% rearrange radiance
R = zeros((M-W+1)*(N-W+1),W*W);
index1 = j+(i-1)*size(Radiance_ext,1);
[u,v] = meshgrid([-win:win],[-win:win]);
index2 = u(transpose(:))'+v(transpose(:))'*size(Radiance_ext,1);
index = index1*ones(1,(2*win+1)*(2*win+1))+...
        ones((M-W+1)*(N-W+1),1)*index2;
R = Radiance_ext(index);

% compute a defocused image for each focus setting
for k=1:K

    ind = ind+size(BlurMap_ext,1)*size(BlurMap_ext,2);
    sigma = BlurMap_ext(ind);
    dirac = sigma<=tolerance;
    isigma = (1-dirac)./(sigma+dirac);
    wsip = isigma(:)*wip4(:)';
    wsim = isigma(:)*wim4(:)';
    derf = (1-exp(-abs(wsip))).*sign(wsip)-...
            (1-exp(-abs(wsim))).*sign(wsim);
    H = .25*derf.*derf(:,transpose(:))+...
        dirac(:)*((wi(:)==0).*(wj(:)==0))';

    if nargin>3
        sigma_dx = dBlurMap_ext(ind);
        dirac = sigma_dx<=tolerance;
        isigma_dx = (1-dirac)./(sigma_dx+dirac);
        wsip = isigma_dx(:)*wip4(:)';
        wsim = isigma_dx(:)*wim4(:)';
        derf = (1-exp(-abs(wsip))).*sign(wsip)-...
                (1-exp(-abs(wsim))).*sign(wsim);
        H_dx = .25*derf.*derf(:,transpose(:))+...
                dirac(:)*((wi(:)==0).*(wj(:)==0))';
    end
```

```
    I(win+1:M-win,win+1:N-win,k) = ...
        reshape(sum(H.*R,2),M-W+1,N-W+1);
    if nargin>3
        I_dx(win+1:M-win,win+1:N-win,k) = ...
            reshape(sum(H_dx.*R,2),M-W+1,N-W+1);
    end
    Kernel(win+1:M-win,win+1:N-win,1:W,1:W) = ...
        reshape(H,M-W+1,N-W+1,W,W);

    if nargin>2
        % compute Lucy-Richardson multiplying factor
        Ratio(win+1:M-win,win+1:N-win) = ...
            I0_ext(win+1:M-win,win+1:N-win,k)./...
            I(win+1:M-win,win+1:N-win,k);
        for w1=1:W
            for w2=1:W
                i = [win+1:N-win]-(w1-win-1);
                j = [win+1:M-win]-(w2-win-1);
                Ker = Kernel(j,i,w1,w2);
                H0(win+1:M-win,win+1:N-win,k) = ...
                    H0(win+1:M-win,win+1:N-win,k)+Ker;
                F0(win+1:M-win,win+1:N-win,k) = ...
                    F0(win+1:M-win,win+1:N-win,k)+Ker.*Ratio(j,i);
            end
        end
    end

end
if nargin>2
    % compute Lucy-Richardson multiplying factor
    HH = sum(H0(win+1:M-win,win+1:N-win,:),3);
    F = sum(F0(win+1:M-win,win+1:N-win,:),3)./HH;
end
if nargin>3
    % restore original domain
    I_dx = I_dx(win+1:M-win,win+1:N-win,:);
end
% restore original domain
I = I(win+1:M-win,win+1:N-win,:);

return

%%%%%%%%%%%%%%%%%%%%%%%%%%%%%%%%%%%%%%%%%%
% testing the approximation of erf
% via the exponential
%%%%%%%%%%%%%%%%%%%%%%%%%%%%%%%%%%%%%%%%%%

x = [-5:.1:5];
Original = erf(x);
Approximate = (1-exp(-abs(x)*1.694)).*sign(x);

figure(1)
```

```
plot(x,Original,'b')
hold on
plot(x,Approximate,'r');
hold off
legend('Original','Approximate');
drawnow
```

E.3   Shape from defocus via diffusion (Chapter 6)

●Shape from Defocus via Diffusion Main Scrip●

```
%
% Depth from Defocus via Diffusion
%

clear all
close all

No = 0;
Yes = 1;

% set the maximum number of iterations
MaxIterations = 100;

fprintf('Shape estimation via relative diffusion\n\n');
% shape type (wave, slope, box, sin, plane, realdata)
notValid = Yes;
while notValid
    shape = input(['What test do you want to perform?'...
        '\nChoose wave, slope, box, sin, plane, realdata: '],'s');
    notValid = ~strcmp(shape,'wave')&...
        ~strcmp(shape,'slope')&...
        ~strcmp(shape,'box')&...
        ~strcmp(shape,'sin')&...
        ~strcmp(shape,'plane')&...
        ~strcmp(shape,'realdata');
end

%%%%%%%%%%%%%%%%%%%%%%%%%%%%%%%%%%%%%%%
% Dataset initialization
%%%%%%%%%%%%%%%%%%%%%%%%%%%%%%%%%%%%%%%
% choose between loading or generating defocused images
LoadImage = strcmp(shape,'realdata');
if LoadImage == No
    % generate synthetic image
    fprintf('Generating synthetic dataset...');
    % set image size
    M = 101; % use odd numbers
```

```
N = 101; % use odd numbers
% lens parameters
F = 12e-3; % focal length
Fnumb = 2; % F number
D = F/Fnumb; % aperture (effective lens diameter)
gamma = 2e4; % calibration parameter
p = 1./(1/F-1./[.52 .85]); % distance CCD-lens
K = length(p); % number of focus settings
% generate depth map
[X,Y] = meshgrid([-(N-1)/2:(N-1)/2],[-(M-1)/2:(M-1)/2]);
switch shape
    case 'wave'
        DepthMap = (1+cos(-(X.^2+Y.^2)/200))./...
            sqrt(X.^2/8600+Y.^2/8600+1)/2*.33+.52;
    case 'slope'
        % along X
        DepthMap = (X-min(X(:)))/(max(X(:))-min(X(:)))*.33+.52;
    case 'box'
        block1 = floor(M/5:M/5*3);
        block2 = floor(M/5*2:M/5*4);
        block3 = floor(N/10:N/2);
        block4 = floor(N/10*4:N/10*9);
        DepthMap = ones(M,N)*.52;
        DepthMap(block1,block3) = ...
            DepthMap(block1,block3)+.2;
        DepthMap(block2,block4) = ...
            DepthMap(block2,block4)+.13;
    case 'sin'
        DepthMap = (sin(X/3)+1)/2*.33+.52;
    otherwise % plane
        DepthMap = .05+ones(M,N)*sum(p)*F/(sum(p)-2*F);
end
% compute true domain
OmegaTruth = double(DepthMap<sum(p)*F/(sum(p)-2*F));
% generate random texture (test algorithm on different intensities)
a1 = 50; a2 = 120; a3 = 255;
Radiance = [a1*rand(floor(M/3),N);...
    a2*rand(floor(M/3),N);...
    a3*rand(M-2*floor(M/3),N)]+1; % skip zero
% add space-varying blurring to test performance of algorithm with
% different textures (high-freq to low-freq)
TextureDepth = p(1)*F/(p(1)-F)*ones(M,N)+...
    [zeros(M,floor(N/3)) ...
    .15*ones(M,floor(N/3)) ...
    .3*ones(M,N-2*floor(N/3))];
% set camera parameters
parms.m = M;
parms.n = N;
parms.F = F;
parms.Fnumb = Fnumb;
parms.D = D;
parms.gamma = gamma;
% same settings for second image
```

```
    parms(2) = parms(1);
    % plane in focus distance (from the lens)
    parms(1).p = p(1);
    parms(2).p = p(2);
    % blur the radiance in three columns
    % compute the relative diffusion map
    Sigma = (gamma*D/2*abs(1-p(1).*(1/F-1./TextureDepth))).^2;
    Radiance = synthesize(Radiance,Sigma);
    % synthesizing blurred images
    Sigma1 = (gamma*D/2*abs(1-p(1).*(1/F-1./DepthMap))).^2;
    Sigma2 = (gamma*D/2*abs(1-p(2).*(1/F-1./DepthMap))).^2;
    I(:,:,1) = synthesize(Radiance,Sigma1);
    I(:,:,2) = synthesize(Radiance,Sigma2);
    % select the domain where the data term is considered
    ww = 3;
    mask = ones(M,N);
    mask(1:ww,:) = 0;
    mask(end-ww+1:end,:) = 0;
    mask(:,1:ww) = 0;
    mask(:,end-ww+1:end) = 0;
    dt_data = 2e-1;% data term step
    dt_reg = 2e-1;% regularization term step
    maxDelta = 1; % max amount of relative blur
    win = 2; % regularization of initial depth map
    fprintf('done\n');
else
    % load images
    fprintf('Loading dataset...');
    load('DataSet.mat');
    % first resize
    scalingFactor = 1/2;
    I1 = imresize(I1,scalingFactor); % limit number of computations
    I2 = imresize(I2,scalingFactor); % limit number of computations
    % select region of interest
    I(:,:,2) = I1(1:end-10,60:end-50);
    I(:,:,1) = I2(1:end-10,60:end-50);
    [M,N,K] = size(I);
    % lens parameters
    F = 12e-3; % focal length
    Fnumb = 2; % F number
    D = F/Fnumb; % aperture (effective lens diameter)
    gamma = 3e4; % calibration parameter
    p = 1./(1/F-1./[.52 .85]); % distance CCD-lens)
    % set camera parameters
    parms.m = M;
    parms.n = N;
    parms.F = F;
    parms.Fnumb = Fnumb;
    parms.D = D;
    parms.gamma = gamma;
    % same settings for second image
    parms(2) = parms(1);
    % plane in focus distance (from the lens)
```

```matlab
    parms(1).p = p(1);
    parms(2).p = p(2);
    % select the domain where the data term is considered
    ww = 3;
    mask = ones(M,N);
    mask(1:ww,:) = 0;
    mask(end-ww+1:end,:) = 0;
    mask(:,1:ww) = 0;
    mask(:,end-ww+1:end) = 0;
    dt_data = 2e-1;%6e-1; % data term step
    dt_reg  = 1e0;%2e0;  % regularization term step
    maxDelta = 1; % max amount of relative blur
    win = 5; % regularization of initial depth map
    K = length(p); % number of focus settings
    fprintf('done\n')
end

% preconditioning threshold
thresh = 6e-2;

% choose whether to use the fast initialization or not
% init type (Y,N)
notValid = Yes;
while notValid
    init = input(['Do you want to use '...
        'the fast initialization (Y/N)?'],'s');
    notValid = ~strcmp(init,'Y')&...
        ~strcmp(init,'N');
end

if strcmp(init,'Y')
    % initialize depth map
    DepthMap_e = initialize(I,maxDelta,win,gamma,F,D,p,'l2');
else
    % initialize depth map with a plane
    DepthMap_e = sum(p)*F/(sum(p)-2*F)*ones(M,N);
end

% compute the relative diffusion map
Sigma1 = (gamma*D/2*abs(1-p(1).*(1/F-1./DepthMap_e))).^2;
Sigma2 = (gamma*D/2*abs(1-p(2).*(1/F-1./DepthMap_e))).^2;
Dsigma = Sigma2-Sigma1;

% start deblurring iterations
fprintf('Shape estimation (%3i iterations)\nIteration       ',...
        MaxIterations);
for i=1:MaxIterations
    fprintf('\b\b\b\b\b[%3i]',i);
    %%%%%%%%%%%%%%%%%%%%%%%%%%%%%%%%%%%%%%%%%
    % Optimization procedure starts here %
    %%%%%%%%%%%%%%%%%%%%%%%%%%%%%%%%%%%%%%%%%
    modelnoise = 0;
    sensornoise = 0;
```

```
Omega = double(Dsigma>0); % Heaviside function
delta = zerocrossing(Dsigma); % Dirac delta
% compute diffusion PDE solution by simulation
[dummy,u1] = synthesize(I(:,:,1),Dsigma.*Omega,...
      modelnoise,sensornoise,5);
[dummy,u2] = synthesize(I(:,:,2),-Dsigma.*(1-Omega),...
      modelnoise,sensornoise,5);
% compute initial conditions for the adjoint PDE
w01 = reshape(u1(:,:,end),M,N)-I(:,:,2);
% compute adjoint diffusion PDE solution by simulation
[dummy,w1] = synthesize(w01,Dsigma.*Omega,...
      modelnoise,sensornoise,5);
w02 = reshape(u2(:,:,end),M,N)-I(:,:,1);
% compute adjoint diffusion PDE solution by simulation
[dummy,w2] = synthesize(w02,-Dsigma.*(1-Omega),...
      modelnoise,sensornoise,5);
%%%%%%%%%%%%%%%%%%%%%%%%%%%%%%%%%%%%%%%%%%%%%
% cost functional gradients computation %
%%%%%%%%%%%%%%%%%%%%%%%%%%%%%%%%%%%%%%%%%%%%%
% time integration (from 0 to T) of grad v grad u
[dvdu1] = timeintegrate(u1,w1);
% time integration (from 0 to T) of grad v grad u
[dvdu2] = timeintegrate(u2,w2);
dEdc = -2*Omega.*dvdu1+2*(1-Omega).*dvdu2+...
       delta.*((u1(:,:,end)-I(:,:,2)).^2-...
       (u2(:,:,end)-I(:,:,1)).^2);
% preconditioning
[dummy,p1] = synthesize(reshape(I(:,:,2),M,N,1),...
            Dsigma.*Omega,modelnoise,sensornoise,5);
[dummy,p2] = synthesize(reshape(I(:,:,1),M,N,1),...
            -Dsigma.*(1-Omega),modelnoise,sensornoise,5);
% time integration (from 0 to T) of grad p grad u
[dpdu1] = timeintegrate(u1,p1);
[dpdu2] = timeintegrate(u2,p2);
precond = 1+abs(-2*Omega.*dpdu1+2*(1-Omega).*dpdu2);
dEdc = dEdc./precond;
dEdc = (abs(dEdc) < thresh).*dEdc+...
       (abs(dEdc) >= thresh).*thresh.*sign(dEdc);
dDatads = dEdc.*mask;
% make sure the number of iterations is an even number
nIt = 2*(floor(dt_reg/1e-1)+1);
beta_data = dt_data/nIt;
beta_regularization = dt_reg/nIt;
for tt=1:nIt
    % compute gradient of the cost functional (data term +
    % regularization) wrt the depth map
    if tt-floor(tt/2)*2==0
        dsigma = beta_data*dDatads - beta_regularization.*...
            laplacian(Dsigma,'forward');
    else
        dsigma = beta_data*dDatads - beta_regularization.*...
            laplacian(Dsigma,'backward');
    end
```

```
        Dsigma = Dsigma-dsigma;
    end
    DepthMap_e = 1./(1/F-(p(2)-p(1)-sqrt((p(2)-p(1))^2+...
                4*(p(2)^2-p(1)^2)*Dsigma/gamma^2/D^2))/...
                (p(2)^2-p(1)^2));
end
fprintf('\ndone\n');

return
```

==

•synthesize•

==

```
function [I,J] = synthesize(radiance,Sigma2,...
    modelnoise,sensornoise,Nit)
% The following code generates synthetic blurred images
% via diffusion

% set image size
if nargin<3
    modelnoise = 0;
end
if nargin<4
    sensornoise = 0;
end
if nargin<5
    Nit = 5;
end
[m,n,three] = size(radiance);

di = .1;
I = zeros(m,n,three);
% search for max number of iterations (Nit)
for k = 1:three
    Sigma2 = Sigma2.*(Sigma2>0);
    DiffCoeff(:,:,k) = Sigma2;
    if (Nit<ceil(max(Sigma2(:))/di))
        Nit = ceil(max(Sigma2(:))/di);
    end
end
if Nit>30
    errmsg = sprintf(...
        ['Current simulation requires %d iterations!\n' ...
        'To proceed change max number of iterations in' ...
        ' function synthesize.m'],Nit);
    error(errmsg);
end
% simulate diffusion
for k = 1:three
    [I(:,:,k),J(:,:,:,k)] = diffusion(radiance(:,:,k)+...
        modelnoise*randn(m,n),DiffCoeff(:,:,k),Nit,...
        sensornoise*randn(m,n));
```

```
end
return
```

•diffusion•

```
function  [I,J] = diffusion(radiance,diffcoeff,Nit,noise)
% Simulate the diffusion equation for Nit steps

J = radiance;
I = radiance;
for i=1:Nit
    % diffuse (one step)
    if mod(i,2)==0
        div = divergence(J(:,:,i),diffcoeff/Nit);
    else
        div = divergenceR(J(:,:,i),diffcoeff/Nit);
    end
    J(:,:,i+1) = J(:,:,i)+div;
end
I = J(:,:,Nit+1)+noise;
return
```

•divergence•

```
function [divF] = divergence(F,coeff)
% compute the divergence of the gradient of f
% multiplied by a diffusion coefficient
% forward difference scheme

[nri,nrj]=size(F);
Gxx = zeros(nri,nrj);
Gyy = zeros(nri,nrj);
Gx(:,2:nrj) = diff(F,1,2);
Gy(2:nri,:) = diff(F,1,1);
% diffusion coefficient
Gxcoeff = Gx.*coeff;
Gycoeff = Gy.*coeff;
Gxx(:,1:nrj-1) = diff(Gxcoeff,1,2);
Gyy(1:nri-1,:) = diff(Gycoeff,1,1);
divF = Gxx+Gyy;
return
```

•divergenceR•

```
function [divF] = divergenceR(F,coeff)
```

```
% compute the divergence of the gradient of f
% multiplied by a diffusion coefficient
% backward difference scheme

[nri,nrj]=size(F);
Gx = zeros(nri,nrj);
Gy = zeros(nri,nrj);
Gx(:,1:nrj-1) = diff(F,1,2);
Gy(1:nri-1,:) = diff(F,1,1);
% diffusion coefficient
Gxcoeff = Gx.*coeff;
Gycoeff = Gy.*coeff;
Gxx(:,2:nrj) = diff(Gxcoeff,1,2);
Gyy(2:nri,:) = diff(Gycoeff,1,1);
divF = Gxx+Gyy;
return
```

●zerocrossing●

```
function [Z] = zerocrossing(I)
% Compute the zero crossings of I

[M,N] = size(I);
J = zeros(M+2,N+2);
J(1+[1:M],1+[1:N],1) = I;
J(1+[1:M],1+[1:N],2) = J([1:M],1+[1:N],1);
J(1+[1:M],1+[1:N],3) = J(2+[1:M],1+[1:N],1);
J(1+[1:M],1+[1:N],4) = J([1:M],[1:N],1);
J(1+[1:M],1+[1:N],5) = J(1+[1:M],[1:N],1);
J(1+[1:M],1+[1:N],6) = J(2+[1:M],[1:N],1);
J(1+[1:M],1+[1:N],7) = J([1:M],2+[1:N],1);
J(1+[1:M],1+[1:N],8) = J(1+[1:M],2+[1:N],1);
J(1+[1:M],1+[1:N],9) = J(2+[1:M],2+[1:N],1);
% positions in the matrix J of the 3x3
% neighbourhood grid
% 4 2 7
% 5 1 8
% 6 3 9
J = J(1+[1:M],1+[1:N],:);
CrossTB = J(:,:,2).*J(:,:,3)<=0;
CrossLTDRBD = J(:,:,4).*J(:,:,9)<=0;
CrossLR = J(:,:,5).*J(:,:,8)<=0;
CrossLBDRTD = J(:,:,6).*J(:,:,7)<=0;
Z = (CrossTB|CrossLTDRBD|CrossLR|CrossLBDRTD);
% check for almost zero
Z = double(Z|(abs(I)<=1e-2));

return
```

●timeintegrate●

```
function  [dvdu] = timeintegrate(J,JJ);
% Integrates \nabla v \cdot \nabla u in time

[m,n,three] = size(J);
dvdu = zeros(m,n);
% time integration (from 0 to T) of grad v grad u
dt = 1/size(J,3);
for t=1:size(J,3)
  jx = zeros(m,n);
  jy = zeros(m,n);
  Jx = zeros(m,n);
  Jy = zeros(m,n);
  if mod(t,2)==0
    Jx(:,2:n) = diff(JJ(:,:,end+1-t),1,2);
    Jy(2:m,:) = diff(JJ(:,:,end+1-t),1,1);
    jx(:,2:n) = diff(J(:,:,t),1,2);
    jy(2:m,:) = diff(J(:,:,t),1,1);
  else
    Jx(:,1:n-1) = diff(JJ(:,:,end+1-t),1,2);
    Jy(1:m-1,:) = diff(JJ(:,:,end+1-t),1,1);
    jx(:,1:n-1) = diff(J(:,:,t),1,2);
    jy(1:m-1,:) = diff(J(:,:,t),1,1);
  end
  dvdu = dvdu+jx.*Jx+jy.*Jy;
end
dvdu = dt*dvdu;
return
```

E.4   Initialization: A fast approximate method

●initialize●

```
function [depth,radiance] = ...
    initialize(I,maxDelta,win,gamma,F,D,p,costType)
% Compute initial depth and radiance with approx but efficient
% method

[M,N,K] = size(I);
%%%%%%%%%%%%%%%%%%%%%%%%%%%%%%%%%%%%%%%%%%%%%%%%%%%%%%%%%%%%%%%%%
% fast initialization (only for 2 defocused images datasets)
%%%%%%%%%%%%%%%%%%%%%%%%%%%%%%%%%%%%%%%%%%%%%%%%%%%%%%%%%%%%%%%%%
fprintf('Initializing depth map...   ');
```

```
NLevels = 50;
deltasigma = [0:maxDelta/(NLevels-1):maxDelta];
E = zeros(M,N,K,length(deltasigma));
W = 2*win+1;
WW = W*W;
Weights = WW*conv2(ones(M,N),ones(W)/W/W,'same');
Cext = zeros(M+2*W,N+2*W,K);
for k=1:size(I,3)
    [J,JJ(:,:,:,k)] = diffusion(I(:,:,k),maxDelta,NLevels-1,0);
end
for i=1:length(deltasigma)
    fprintf('\b\b\b\b[%2i]',i)
    J = squeeze(JJ(:,:,i,:));
    % choose either the l2 norm or the I-divergence
    if strcmp(costType,'l2')
        Cost = (I(:,:,end:-1:1)-J).^2;
    else
        Cost = I(:,:,end:-1:1).*log(I(:,:,end:-1:1)./J)-...
            I(:,:,end:-1:1)+J;
    end
    %%%%%%%%%%%%%%%%%%%%%%%%%%%%%%
    % original implementation
    %Cost = convn(Cost,ones(W)/WW,'same');
    %E(:,:,:,i) = Cost./repmat(Weights,[1 1 K]);
    %%%%%%%%%%%%%%%%%%%%%%%%%%%%%%%%
    % alternative implementation
    % use integral images instead of doing convolution
    % this is faster when the kernel is rectangular and large
    Cext(W+[1:M],W+[1:N],:) = Cost;
    for k=1:K
        % compute integral image
        CC = cumsum(cumsum(Cext(:,:,k),1),2);
        E(:,:,k,i) = (CC(2*W-win:end-W+win,2*W-win:end-W+win)+...
            CC(W-win:end-2*W+win,W-win:end-2*W+win)-...
            CC(2*W-win:end-W+win,W-win:end-2*W+win)-...
            CC(W-win:end-2*W+win,2*W-win:end-W+win))./Weights;
    end
end
% search for minimum energy
for k=1:K
    [nrg(:,:,k),dd(:,:,k)] = min(E(:,:,k,:),[],4);
end
[dummy,kk] = min(nrg,[],3);
ind = zeros(M,N);
for k=1:K
    ind = ind+dd(:,:,k).*(kk==k);
end
sgn = (kk==1)-(kk==2);
% correct for discrepancies due to the integral images
sigma2 = .93*deltasigma(ind).*sgn;
% additional smoothing
for t=1:10
    if mod(t,2)==0
```

```
        sigma2 = sigma2 + 1e-1*laplacian(sigma2,'forward');
    else
        sigma2 = sigma2 + 1e-1*laplacian(sigma2,'backward');
    end
end
a = p(1)^2-p(2)^2;
b = -2*(p(1)-p(2));
c = sigma2/gamma^2/(D/2)^2;
fprintf('done\n');

% initialize depth map
depth = 1./(1/F-(-b+sqrt(abs(b^2-4*a.*c)))/2/a);
% pick the sharpest regions and eliminate zeroes
radiance = I(:,:,2).*(kk==2)+I(:,:,1).*(kk==1);

return
```

Appendix F
Regularization

In this appendix we give a rough sketch of the basics of regularization theory, which is used throughout the book in order to solve ill-posed inverse problems that arise while inferring shape and radiance from images. We first define inverse (as opposed to direct) problems, then describe ill-posed (as opposed to well-posed) problems, and finally describe regularization theory. In particular, we show how the two forms of regularization that we have employed, namely the addition of constraints to the energy functional (Chapters 5) and the truncation of the functional SVD (Chapters 4), are related.

Before this appendix, the reader may want to review Section 2.1.4 where the operatorial notation is introduced. The reader who is interested in a more germane treatment of the topic of regularization should consult the appropriate textbooks, for instance [Kirsch, 1996]. The reader may also want to revisit Appendix B, where the functional derivative is defined.

F.1 Inverse problems

The concept of an *inverse problem* finds a first intuitive definition in the context of physics, where the notion of cause and that of effect is related to a natural phenomenon. A direct problem is that of determining the effect given the cause, whereas the inverse problem is that of determining the cause given the effect. To bring this notion closer to our context, consider the basic model of the image formation process described in Section 2.1.4,

$$I = Hr, \tag{F.1}$$

where the scene, with its radiance r and shape encoded in H, causes the formation of an image I. Equation (F.1) describes the direct problem. Given r and H, it tells us precisely how to compute I. The inverse problem, on the other hand, is that of recovering r (as well as H, but let us assume for a moment that H is known, for simplicity) given I.

Note that equation (F.1) does not tell us how to solve the inverse problem. This may be very easy, for instance if H is an invertible operator (say, for instance, a square invertible matrix), or it may be impossible (think of a singular matrix, for instance, a tall and skinny one with full column rank). Even more problematic is the case when r is a function, that is, an element of an infinite-dimensional space, whereas I is finite-dimensional, such as the case of digital images.

Indeed, it is common for the cause (e.g., the scene) to be more complex, in the sense of belonging to a higher-dimensional space, than the effects; one could say that "information is lost" in the direct problem. This makes the inverse problem particularly difficult because one has to "re-create information" lost in the data generation process. In particular, there may be many, indeed infinitely many, different causes r that yield the same effect I, so how can we choose one cause? Is the attempt to find a cause even meaningful?

Whether the solution of a problem is meaningful is a question that occupied Hadamard at the turn of the century, and led to the definition of ill-posed problems, which we discuss in the next section, and the ensuing theory of regularization, which we discuss in the following one.

Example (Inverse diffusion).   Consider the model (F.1), where H represents convolution by a Gaussian kernel, as in Section 2.1.4. This can be equivalently described by a partial differential equation, the diffusion equation, as we showed in Section 2.4. Thus, the direct problem consists of taking a function (e.g., a radiance) r and "blurring" it. Clearly, many details of the function are lost in the process, that eventually converges to a constant.

The inverse problem, that of recovering the original function r from a blurred version I, is a typical example of ill-posed problem. Before we give a formal definition, in the next section, we show with an example what we mean by a "loss of information."

Suppose that we are given the initial radiance or, to continue the heat analogy introduced in Section 2.4, a temperature distribution $r : \mathbb{R} \mapsto [0, \infty)$ at time $t = 0$,

$$r(x) = \sum_{i=1}^{\infty} r_i \sin(\pi i x), \tag{F.2}$$

where

$$r_i = 2 \int_0^1 r(x) \sin(\pi i x) dx \tag{F.3}$$

and that heat diffuses uniformly with diffusion coefficient $c > 0$, that is,

$$\begin{cases} \dot{u}(x,t) & = & c\frac{\partial^2 u(x,t)}{\partial x^2} \\ u(x,0) & = & r(x) \\ u(0,t) & = & u(1,t) = 0. \end{cases} \tag{F.4}$$

If we are given the initial temperature distribution r with precision ϵ; that is, we can only observe r_i such that $|r_i| > \epsilon$, then, because $r_{i+1} \leq r_i$,

$$r(x) = \sum_{i=1}^{N} r_i \sin(\pi i x) \tag{F.5}$$

for a finite integer $N > 0$. Now, because the solution of equation (F.4) is

$$u(x,t) = \sum_{i=1}^{\infty} r_i e^{-(\pi i)^2 tc} \sin(\pi i x) \tag{F.6}$$

we notice that due to the quick decay of the factor $e^{-(\pi i)^2 tc}$, the terms satisfying $|r_i e^{-(\pi i)^2 tc} \sin(\pi i x)| = |r_i e^{-(\pi i)^2 tc}| < \epsilon$ are fewer than N. This loss in the number of coefficients means that the initial knowledge about the temperature distribution r has been lost during the diffusion process.

The inverse problem in this example amounts to recovering the initial temperature distribution r given the temperature $u(x,t)$ measured at time t. This requires recovering a number of coefficients that have been lost during the diffusion process.

F.2   Ill-posed problems

The basic definition of a well-posed problem was introduced by [Hadamard, 1902] in order to formalize the notion of a meaningful solution mentioned in the previous section. The complete formulation of the theory of ill-posed problems is due to [R.Courant and Hilbert, 1962]. In short, a problem is well-posed when:

1. The solution exists for any data

2. The solution is unique

3. The solution depends continuously on the data.

Clearly, if the solution exists for some set of data but not others, or if there are many solutions, or if a slight perturbation of the data causes a drastic change in the solution, such a solution is not very meaningful. Whenever any of the requirements above are not satisfied, the problem is said to be ill-posed.

In our case, the first requirement is always satisfied given that, no matter how noisy, measured images have finite energy, and there always exists (at least) one scene that generates it. The main issues that we need to deal with in our shape estimation and image restoration problem is to enforce the uniqueness of the solution and to make sure that small changes in the image (for instance, due to noise) do not cause major changes in the solution. This is done through regularization, which we describe next.

F.3 Regularization

In general terms, regularization refers to the approximation of an ill-posed problem by a collection of well-posed problems that are "close" to the original one, and converge to it in a sense that we make clear shortly. The general theory was laid out by A. N. Tikhonov, although today "Tikhonov regularization" refers to a special case of least-squares problem. In the context of the linear inverse problems that have (F.1) as their direct version, the main idea is that we are given noisy data I^δ where δ denotes some perturbation (this could be additive noise, as in our case, or a more general perturbation, so we leave the notation generic at this point) such that

$$\left\| Hr^\delta - I^\delta \right\| < \delta. \tag{F.7}$$

Then the approximate solution r^δ, that must depend continuously on the data, can be computed by constructing a linear and bounded operator, call it R, so that

$$r^\delta = RI^\delta. \tag{F.8}$$

The operator R can be obtained as the limit of a family of bounded and linear operators R_α, $\alpha > 0$ such that

$$\lim_{\alpha \leftarrow 0} R_\alpha Hr = r \quad \forall r. \tag{F.9}$$

This is the requirement of a meaningful solution: when the parameter α tends to 0, the solution tends to the solution of the unperturbed problem. Now, let I be the unperturbed data in the range of H, and I^δ be the noisy data such that $\| I - I^\delta \| \leq \delta$. If we define

$$r^{\alpha,\delta} = R_\alpha I^\delta \tag{F.10}$$

as a family of approximate solutions to (F.1), then the discrepancy between our approximate solution and the unperturbed solution can be split into two parts:

$$\begin{aligned} \| r^{\alpha,\delta} - r \| &\leq \| R_\alpha I^\delta - R_\alpha I \| + \| R_\alpha I - r \| \\ &\leq \| R_\alpha \| \| I^\delta - I \| + \| R_\alpha Hr - r \| \\ &\leq \delta \| R_\alpha \| + \| R_\alpha Hr - r \|, \end{aligned} \tag{F.11}$$

where the first term of the right-hand side is the noise-propagation error multiplied by the "condition number" $\| R_\alpha \|$ of the regularized problem. The second term is the approximation error. Whereas the noise-propagation error tends to infinity with α going to zero and to zero when α goes to infinity, the approximation error goes to zero when α goes to infinity and to infinity when α goes to zero.

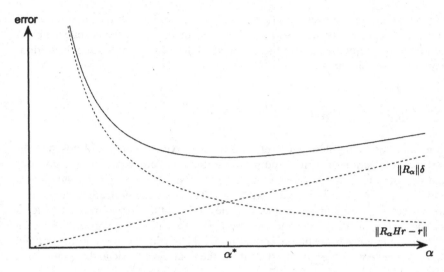

Figure F.1. Behavior of the total error. Whereas the noise-propagation error tends to infinity with α going to zero and to zero when α goes to infinity, the approximation error goes to zero when α goes to infinity and to infinity when α goes to zero.

The behavior is illustrated in Figure F.1. Because the discrepancy between our approximate solution and the unperturbed solution is a combination of the two, as shown in equation (F.11), and given their general behavior described above, there is a unique parameter α^* that minimizes equation (F.11). It is part of a minimization strategy to find such optimal α^*. In our analysis it is important to know that such a parameter exists and is unique.

In summary, regularization can be defined as a general strategy to search for families of approximate solutions satisfying additional constraints coming from the physics of the problem. For instance, one could require the solution to have bounded energy or intensity (Chapters 4–9), to be positive (Chapters 5), or be smooth (see Chapters 4–9), or be subject to statistical constraints on the noise distribution (Chapter 2).

Remark F.1. Regularization versus post processing
In many applications it is common to "clean up" the solution after it has been obtained. This process is often called "post processing." Post processing is a procedure where one applies the constraints above (e.g., positivity, smoothness) after having obtained some nominal solution. From the analysis above, it should be clear that post processing is not a sensible method for ill-posed problems; although it may decrease the noise level, it does so at the cost of increasing the approximation error, and due to the non continuous dependency on the data perturbation, it may do so in an unpredictable way.

F.3.1 Tikhonov regularization

The term "Tikhonov regularization" often refers to a special case of the procedure described above to determine the smoothest approximate solution of a constrained least-squares problem, as developed by Phillips [Phillips, 1962].

In this section we illustrate how Tikhonov regularization applies to our specific inverse problems, and how this relates to the truncated singular value decomposition presented in Section 4.2.1,

$$E_\alpha(r) = \|Hr - I^\delta\|^2 + \alpha\|r\|^2, \qquad (F.12)$$

where α is a constant.[1] If we seek to minimize equation (F.12) given the perturbed image I^δ, then we obtain the following Euler equation (see Appendix B)

$$(H^*H + \alpha I_d)r^\alpha = H^*I^\delta, \qquad (F.13)$$

where I_d denotes the identity operator and $(\cdot)^*$ denotes the adjoint operator, defined in Section 2.1.4, equation (2.15). Then, the approximate solution $r^{\alpha,\delta}$ can be written as

$$r^{\alpha,\delta} = R_\alpha I^\delta, \qquad (F.14)$$

where

$$R_\alpha = (H^*H + \alpha I_d)^{-1}H^*. \qquad (F.15)$$

As one can see, as α tends to zero, the approximate solution in equation (F.14) tends to the least-squares solution which we presented in Chapter 4, equation (4.26). Now, suppose that the perturbed image satisfies the following model

$$I^\delta = Hr + w^\delta, \qquad (F.16)$$

where w^δ is a perturbation. As we have seen before, the error between our solution and the unperturbed problem solution is given by

$$\|R_\alpha I^\delta - R_0 I\| \le \|(R_\alpha H - R_0)r\| + \|R_\alpha w^\delta\|, \qquad (F.17)$$

where the two terms on the right-hand side are the approximation error and the noise-propagation error. If we use the singular value decomposition (see Chapter 4), we obtain that

$$H = \sum_{i=0}^{\infty} \lambda_i u_i v_i, \qquad (F.18)$$

where λ_i is a sequence of nonnegative scalars, v_i is an orthonormal set of vectors (because we consider that image to be a digital image, I is a vector), and u_i is an

---

[1]Those familiar with constrained optimization will recognize α as being the Lagrange multiplier, part of the unknowns in the minimization of the energy in (F.12). Those interested in the many details we are glancing over here can consult [Polak, 1997], Chapter X.

orthonormal set of functions in L^2. Then,

$$r^{\alpha,\delta} = R_\alpha I^\delta = \sum_{i=0}^{\infty} \frac{\lambda_i}{\lambda_i^2 + \alpha} u_i \langle v_i, I^\delta \rangle \tag{F.19}$$

and we obtain that the squared approximation error is

$$\|(R_\alpha H - R_0)r\|^2 = \sum_{i=0}^{\infty} \frac{\alpha^2}{(\lambda_i^2 + \alpha)^2} |\langle u_i, r \rangle|^2 \tag{F.20}$$

and the noise-propagation error is

$$\|R_\alpha w^\delta\|^2 = \sum_{i=0}^{\infty} \frac{\lambda_i^2}{(\lambda_i^2 + \alpha)^2} |\langle v_i, w^\delta \rangle|^2. \tag{F.21}$$

Now we can observe that the behavior of equations (F.20) and (F.21) as α tends to 0 or ∞ reflects the definitions given in the previous section.

F.3.2   Truncated SVD

In this section we show that the truncated singular value decomposition used in Chapter 4 is a type of Tikhonov regularization. To see that, we recast the representation of the regularized solution (F.19) in the following form,

$$r^{\alpha,\delta} = \sum_{i=0}^{\infty} \frac{g_{\alpha,i}}{\lambda_i} u_i \langle v_i, I \rangle, \tag{F.22}$$

where $g_{\alpha,i}$ is a *windowing* function defined as

$$g_{\alpha,i} = \frac{\lambda_i^2}{\lambda_i^2 + \alpha}. \tag{F.23}$$

This shows that the regularized solution is a filtered version of the singular value decomposition of the generalized solution. Indeed, if we set

$$g_{\alpha,i} = \begin{cases} 1 & \text{if } \lambda_i^2 \geq \alpha \\ 0 & \text{otherwise} \end{cases} \tag{F.24}$$

we obtain the truncated SVD also employed in Chapter 4. This regularization method is also called *spectral-cutoff*.

References

Ahmed, A. and Farag, A. (2006). A new formulation for shape from shading for non-lambertian surfaces. In *Proc. IEEE Conf. on Comp. Vision and Pattern Recogn.*, volume 2, pages 1817–24.

Akaike, H. (1976). Canonical correlation analysis of time series and the use of an information criterion. *System identification: advances and case studies*, R. Mehra and D. Lainiotis (Eds.):27–96.

Ancin, H., Roysam, B., Dufresne, T., Chestnut, M., and Ridder, G. (1996). Advances in automated 3-d image analysis of cell populations imaged by confocal microscopy. *Cytometry*, pages 221–34.

Asada, N., Fujiwara, H., and Matsuyama, T. (1998a). Edge and depth from focus. *Intl. J. of Comp. Vision*, 26(2).

Asada, N., Fujiwara, H., and Matsuyama, T. (1998b). Seeing behind the scene: Analysis of photometric properties of occluding edges by reversed projection blurring model. *IEEE Trans. Pattern Analysis and Machine Intelligence*, 20:155–67.

Belhumeur, P. and Kriegman, D. (1998). What is the set of images of an object under all possible illumination conditions? *Int. J. of Computer Vision*, 28 (3).

Ben-Ezra, M. and Nayar, S. (2003). Motion deblurring using hybrid imaging. In *Computer Vision and Pattern Recognition*, volume 1, pages 657–664.

Bertero, M. and Boccacci, P. (1998). Introduction to inverse problems in imaging. *Institute of Physics Publishing, Bristol and Philadelphia*.

240   References

Bhat, D. N. and Nayar, S. K. (1995). Stereo in the presence of specular reflection. In *ICCV*, pages 1086 – 1092.

Blake, A. (1985). Specular stereo. In *IJCAI*, pages 973–976.

Borman, S. and Stevenson, R. L. (1998). Super-Resolution from Image Sequences - A Review. In *Proceedings of the 1998 Midwest Symposium on Circuits and Systems*, Notre Dame, IN.

Born, M. and Wolf, E. (1980). Principles of optics. In *Pergamon Press.*

Brelstaff, G. and Blake, A. (1988). Detecting specular reflections using lambertian constraints. In *ICCV*, pages 297–302.

Brostow, G. J. and Essa, I. (2001). Image-based motion blur for stop motion animation. In Fiume, E., editor, *SIGGRAPH 2001, Computer Graphics Proceedings*, pages 561–566.

Chan, T. and Vese, L. (2001). Active contours without edges. *IEEE Transactions on Image Processing*, 10(2):266–77.

Chaudhuri, S. and Rajagopalan, A. (1999). *Depth from defocus: a real aperture imaging approach.* Springer Verlag.

Chen, W., Nandhakumar, N., and Martin, W. (1996). Image motion estimation from motion smear: A new computational model. *PAMI*, 18(4):412–425.

Cheng, Y., Johnson, A., and Matthies, L. (2005). Mer-dimes: A planetary landing application of computer vision. In *CVPR (1)*, pages 806–813.

Cipolla, R. and Blake, A. (1992). Surface shape from the deformation of apparent contours. *Int. J. of Computer Vision, 9 (2).*

Cover, T. M. and Thomas, J. A. (1991). Elements of information theory. *Wiley Interscience.*

Csiszár, I. (1991). Why least squares and maximum entropy? -an axiomatic approach to inverse problems. In *Ann. of Stat.*, volume 19, pages 2033–66.

Dickmanns, E. D. and Graefe, V. (1988). Applications of dynamic monocular machine vision. *Machine Vision and Applications*, 1:241–261.

Engl, H., Hanke, M., and Neubauer, A. (1996). *Regularization of Inverse Problems.* Kluwer Academic Publishers, Dordrecht.

Ens, J. and Lawrence, P. (1993). An investigation of methods for determining depth from focus. *IEEE Trans. Pattern Anal. Mach. Intell.*, 15:97–108.

Evans, L. C. (1992). *Partial differential equations.* American Mathematical Society.

Favaro, P., Burger, M., and Soatto, S. (2004). Scene and motion reconstruction from defocused and motion-blurred images via anisotropic diffusion. In *Proc. of the Eur. Conf. on Comp. Vision*, pages 257–269.

Favaro, P., Mennucci, A., and Soatto, S. (2003). Observing shape from defocused images. *IJCV*, 52(1):25–43.

Favaro, P. and Soatto, S. (2000). Shape and radiance estimation from the information divergence of blurred images. *Proc. IEEE Computer Vision and Pattern Recognition*, pages 755–68.

Favaro, P. and Soatto, S. (2002). Learning shape from defocus. In *Proc. of the Eur. Conf. on Computer Vision (ECCV)*, volume 2, pages 735–745.

Favaro, P. and Soatto, S. (2003). Seeing beyond occlusions (and other marvels of a finite lens aperture). In *Proc. IEEE Conf. on Comp. Vision and Pattern Recogn.*, pages II–579–586.

Favaro, P. and Soatto, S. (2004). A variational approach to scene reconstruction and image segmentation from motion-blur cues. In *CVPR (1)*, pages 631–637.

Favaro, P. and Soatto, S. (2005). A geometric approach to shape from defocus. *IEEE Trans. Pattern Anal. Mach. Intell.*, 27(3):406–417.

Felleman, D. J. and van Essen, D. C. (1991). Distributed hierarchical processing in the primate cerebral cortex. *Cerebral Cortex*, 1:1–47.

Flitcroft, D., Judge, S., and Morley, J. (1992). Binocular interactions in accommodation control: effects of anisotropic stimul. *Journal of Neuroscience*, 12(1):188–203.

Flitcroft, D. and Morley, J. (1997). Accommodation in binocular contour rivalry. *Vision Research*, 37(1):121–125.

Fomin, S. V. and Gelfand, I. M. (2000). *Calculus of Variations*. Dover Publications.

Forsyth, D. A. (2002). Shape from texture without boundaries. In *ECCV (3)*, pages 225–239.

Forsyth, D. A. and Ponce, J. (2003). Computer vision: A modern approach. In *Prentice-Hall*.

Gilardi, G. (1992). Analisi due. In *McGraw-Hill*.

Gokstorp, M. (1994). Computing depth from out-of-focus blur using a local frequency representation. In *International Conference on Pattern Recognition*, pages A:153–158.

Golub, G. and Pereyra, V. (1973). The differentiation of pseudo-inverses and nonlinear least-squares problems whose variables separate. *SIAM J. Numer. Anal.*, 10(2):413–532.

Goodman, J. W. (1996). Introduction to fourier optics. *McGraw-Hill*, 2nd Edition.

Hadamard, J. (1902).   Sur les problémes aux dérivées partielles et leur signification physique. In *Princeton University Bulletin*, pages 49–52.

Hammett, S., Georgeson, M., and Gorea, A. (1998).   Motion blur and motion sharpening: temporal smear and local contrast non-linearity. In *Vision Research*, volume 38, pages 2099–2108.

Hopkins, H. H. (1955).   The frequency response of a defocused optical system. *Proc. R. Soc. London Ser. A*, 231:91–103.

Horn, B. K. P. (1987). Robot vision. In *MIT Press*.

Horn, B. K. P. and Brooks, M. J. (1989).   *Shape from Shading*.   MIT Press, Cambridge Massachusetts.

Howard, I. and Rogers, B. (1995).   Binocular vision and stereopsis.   *Oxford University Press*.

Huang, W. J. and Mumford, D. (1999).   Statistics of natural images and models. In *Computer Vision and Pattern Recognition*, volume 1, pages 541–7.

Ikeuchi, K. (1981).   Determining surface orientations of specular surfaces by using the photometric stereo method.   *IEEE Trans. on Pattern Analysis and Machine Intelligence*,, 3:661–669.

Jin, H. and Favaro, P. (2002).   A variational approach to shape from defocus. In *ECCV02*, page II: 18 ff.

Jin, H., Favaro, P., and Cipolla, R. (2005).   Visual tracking in the presence of motion blur. In *CVPR (2)*, pages 18–25.

Jin, H., Soatto, S., and Yezzi, A. J. (2003a). Multi-view stereo beyond lambert. In *Proc. IEEE Conf. on Comp. Vision and Pattern Recogn.*, pages I–171–178.

Jin, H., Yezzi, A., and Soatto, S. (June 2000). Stereoscopic shading: integrating shape cues in a variational framework. In *Intl. Conf. on Computer Vision and Pattern Recognition*, pages 169–176.

Jin, H. L., Yezzi, A. J., Tsai, Y. H., Cheng, L. T., and Soatto, S. (2003b). Estimation of 3D surface shape and smooth radiance from 2D images: A level set approach. *J. Scientific Computing*, 19:267–292.

Kang, S., Choung, Y., and Paik, J. (1999). Segmentation-based image restoration for multiple moving objects with different motions. In *ICIP99*, pages I:376–380.

Kang, S., Min, J., and Paik, J. (2001).   Segmentation-based spatially adaptive motion blur removal and its application to surveillance systems. In *ICIP01*, pages I: 245–248.

Kim, J., Tsai, A., Cetin, M., and Willsky, A. (2002).   A curve evolution-based variational approach to simultaneous image restoration and segmentation. In *ICIP02*, pages I: 109–112.

Kirsch, A. (1996). *An introduction to the mathematical theory of inverse problems.* Springer Verlag.

Kotulak, J. and Morse, S. (1994). Relationship among accommodation, focus and resolution with optical instruments. *Journal of the Opt. Soc. Am. A,* 11(1):71–79.

Kubota, A. and Aizawa, K. (2002). Arbitrary view and focus image generation: rendering object-based shifting and focussing effect by linear filtering. In *ICIP02*, pages I: 489–492.

Levoy, M., Chen, B., Vaish, V., Horowitz, M., McDowall, I., and Bolas, M. T. (2004). Synthetic aperture confocal imaging. *ACM Trans. Graph.,* 23(3):825–834.

Levoy, M., Ng, R., Adams, A., Footer, M., and Horowitz, M. (2006). Light field microscopy. *ACM Trans. Graph.,* 25(3):924–934.

Liu, J. S. (2001). *Monte Carlo Strategies in Scientific Computing.* Springer-Verlag.

Luenberger, D. (1968). Optimization by vector space methods. *Wiley.*

Ma, J. and Olsen, S. I. (1990). Depth from zooming. In *JOSA A,* volume 7, pages 1883–90.

Ma, Y., Soatto, S., Kosecka, J., and Sastry, S. (2003). *An invitation to 3D vision, from images to models.* Springer Verlag.

Marshall, J., Burbeck, C., and Ariely, D. (1996). Occlusion edge blur: a cue to relative visual depth. *Intl. J. Opt. Soc. Am. A,* 13:681–688.

McOwen, R. (1996). Partial differential equations, methods and applications. In *Prentice-Hall.*

Munkres, J. (1999). *Topology.* Prentice Hall.

Nayar, S. K., Fang, X. S., and Boult, T. (1993). Removal of specularities using color and polarization. In *CVPR,* pages 585–590.

Nayar, S. K. and Nakagawa, Y. (1990). Shape from focus: An effective approach for rough surfaces. In *IEEE International Conference on Robotics and Automation,* pages 218–225.

Noguchi, M. and Nayar, S. K. (1994). Microscopic shape from focus using active illumination. In *International Conference on Pattern Recognition,* pages 147–152.

Okutomi, M. and Kanade, T. (1993). A multiple baseline stereo. *IEEE Trans. Pattern Anal. Mach. Intell.,* 15(4):353–363.

Oppenheim, A. V. and Schafer, R. W. (1975). *Digital Signal Processing.* Prentice Hall.

Pentland, A. (1987). A new sense for depth of field. *IEEE Trans. Pattern Anal. Mach. Intell.*, 9:523–531.

Pentland, A., Darrell, T., Turk, M., and Huang, W. (1989). A simple, real-time range camera. In *Computer Vision and Pattern Recognition*, pages 256–261.

Pentland, A., Scherock, S., Darrell, T., and Girod, B. (1994). Simple range cameras based on focal error. *Journal of the Optical Society of America A*, 11(11):2925–2934.

Phillips, D. L. (1962). A technique for the numerical solution of certain integral equations of the first kind. *J. ACM*, 9(1):84–97.

Polak, E. (1997). *Optimization*. Springer Verlag.

Prados, E. and Faugeras, O. (2003). Perspective shape from shading and viscosity solutions. In *Proc. 9th Int. Conf. on Computer Vision*.

Rajagopalan, A. N. and Chaudhuri, S. (1997). Optimal selection of camera parameters for recovery of depth from defocused images. In *Computer Vision and Pattern Recognition*, pages 219–224.

Rajagopalan, A. N. and Chaudhuri, S. (1998). Optimal recovery of depth from defocused images using an mrf model. In *Proc. International Conference on Computer Vision*, pages 1047–1052.

Rajagopalan, A. N., Chaudhuri, S., and Mudenagudi, U. (2004). Depth estimation and image restoration using defocused stereo pairs. *IEEE Trans. Pattern Anal. Mach. Intell.*, 26(11):1521–1525.

Rav-Acha, A. and Peleg, S. (Palm-Springs, 2000). Restoration of multiple images with motion blur in different directions. In *IEEE Workshop on Applications of Computer Vision (WACV)*, pages 22–28.

R.Courant and Hilbert, D. (1962). *Methods of mathematical physics*. Princeton University Bulletin.

Rissanen, J. (1978). Modeling by shortest data description. *Automatica*, 14:465–471.

Schechner, Y. and Kiryati, N. (1993). The optimal axial interval in estimating depth from defocus. In *Proc. of the Intl. Conf. of Comp. Vision*, pages 843–848.

Schneider, G., Heit, B., Honig, J., and Bremont, J. (1994). Monocular depth perception by evaluation of the blur in defocused images. In *Proc. International Conference on Image Processing*, volume 2, pages 116–9.

Sethian, J. A. (1996). Level set methods: Evolving interfaces in geometry, fluid mechanics, computer vision, and material science. *Cambridge University Press*.

Snyder, D. L., Schulz, T. J., and O'Sullivan, J. A. (1992). Deblurring subject to nonnegativity constraints. *IEEE Transactions on Signal Processing*, 40(5):1142–1150.

Soatto, S. and Favaro, P. (2000). A geometric approach to blind deconvolution with application to shape from defocus. *Proc. IEEE Computer Vision and Pattern Recognition*, 2:10–7.

Soatto, S., Yezzi, A. J., and Jin, H. (2003). Tales of shape and radiance in multiview stereo. In *Proc. of Intl. Conf. on Computer Vision*, volume 1, pages 171–178.

Subbarao, M. (1988). Parallel depth recovery by changing camera parameters. In *Proc. International Conference on Computer Vision*, pages 149–155.

Subbarao, M. and Gurumoorthy, N. (1988). Depth recovery from blurred edges. In *Computer Vision and Pattern Recognition*, pages 498–503.

Taylor, M. (1996). Partial differential equations (volume i: Basic theory). *Springer Verlag*.

Tschumperle, D. and Deriche, R. (2003). Vector-valued image regularization with pde's: A common framework for different applications. In *CVPR03*, pages I: 651–656.

Tull, D. and Katsaggelos, A. (1996). Regularized blur-assisted displacement field estimation. In *ICIP96*, pages III: 85–88.

Vese, L. A. and Chan, T. F. (2002). A multiphase level set framework for image segmentation using the mumford and shah model. *Int. J. of Computer Vision*, 50(3):271–293.

Walsh, G. and Charman, W. (1988). Visual sensitivity to temporal change in focus and its relevance to the accommodation response. *Vision Research*, 28(11):1207–1221.

Watanabe, M. and Nayar, S. (1998). Rational filters for passive depth from defocus. *Intl. J. of Comp. Vision*, 27(3).

Watanabe, M. and Nayar, S. K. (1996a). Minimal operator set for passive depth from defocus. In *Computer Vision and Pattern Recognition*, pages 431–438.

Watanabe, M. and Nayar, S. K. (1996b). Telecentric optics for computational vision. In *European Conference on Computer Vision*, pages II:439–451.

Watanabe, M. and Nayar, S. K. (1997). Telecentric optics for focus analysis. *IEEE Transactions on Pattern Analysis and Machine Intelligence*, 19(12):1360–1365.

web link (2006a). http://www.2d3.com.

web link (2006b). http://www.photomodeler.com.

Weickert, J., Haar, B., and Viergever, R. (1998). Efficient and reliable schemes for nonlinear diffusion filtering. *IEEE Transactions on Image Processing*, 7(3):398–410.

Xiong, Y. and Shafer, S. (1993). Depth from focusing and defocusing. In *Proc. of the Intl. Conf. of Comp. Vision and Pat. Recogn.*, pages 68–73.

Xiong, Y. and Shafer, S. A. (1995). Moment filters for high precision computation of focus and stereo. *Proc. of International Conference on Intelligent Robots and Systems*, pages 108–113.

Yezzi, A. and Soatto, S. (2003). Stereoscopic segmentation. *Intl. J. of Computer Vision*, 53(1):31–43.

Yitzhaky, Y., Mor, I., Lantzman, A., and Kopeika, N. (1998). Direct method for restoration of motion blurred images. *JOSA-A*, 15(6):1512–1519.

You, Y. and Kaveh, M. (1996). Anisotropic blind image restoration. *Proceedings of 3rd IEEE International Conference on Image Processing*, 2:461–4.

You, Y. and Kaveh, M. (1999). Blind image restoration by anisotropic diffusion. *IEEE Trans. on Image Processing*, 8(3):396–407.

Yu, T., Xu, N., and Ahuja, N. (2004). Recovering shape and reflectance model of non-lambertian objects from multiple views. In *CVPR (2)*, pages 226–233.

Yuille, A., Coughlan, J. M., and Konishi, S. (2003). Kgbr viewpoint-lighting ambiguity. *J. Opt. Soc. Am. A*, 20(1):24–31.

Ziou, D. and Deschenes, F. (2001). Depth from defocus estimation in spatial domain. *CVIU*, 81(2):143–165.

Zomet, A., Rav-Acha, A., and Peleg, S. (2001). Robust super-resolution. In *CVPR01*, pages I:645–650.

Index